CW00530714

AQUATIC PLANTS FOR THE WASTE WATER TREATMENT

AQUATIC PLANTS FOR THE WASTE WATER TREATMENT

Alka Rani Upadhyay

Pollution Ecology Research Laboratory
Centre of Advanced Study in Botany
Banaras Hindu University
Varanasi-221005 (INDIA)

2004

DAYA PUBLISHING HOUSE
Delhi - 110035

ISBN 81-7035-335-1

Published by	:	**Daya Publishing House** 1123/74, Deva Ram Park Tri Nagar, Delhi-110 035 Phone : 27383999 Fax : 011-27382329 e-mail : dayabooks@vsnl.com website : www.dayabooks.com
Showroom	:	4762-63/23, Ansari Road, Darya Ganj, New Delhi-110002 Phone : 23245578, 23244987
Laser Typesetting	:	**P. K. Jain & Co.**
Printed at	:	**Chawla Offset Printers** Delhi-110 052

PRINTED IN INDIA

ACKNOWLEDGEMENTS

I will everlastingly remain indebted for the benignity of Almighty, who showed me the path to through.

I consider it to be a rare opportunity of my life to thank my supervisor, Dr. B.D. Tripathi, Co-ordinator, Environmental Science, CAP in Botany, Banaras Hindu University, Varanasi, whose inspiring guidance, keen interest and constructive criticism throughout the course of investigation attributed to the completion of my research work. His sincerity and punctuality has played a role of major catalyst in promoting the onward march of this work at every step. His invigarating encouragement and right direction at every crucial juncture are also appreciable.

I wish to express my gratitude to Prof. V.P. Jaiswal, Head, Department of Botany, Banaras Hindu University, Varanasi for his constant help and cooperation.

Candidly, I am highly obliged to Prof. R.P. Ambasht, Emeritus Scientist CPIR, CAP in Botany, Banaras Hindu University, Varanasi for constant inspiration and critical suggestions.

I do not find adequate words to express my gratitude to my esteemed teacher Dr. B.D. Nautiryal, Department of Botany, University of Lucknow, Lucknow for his suggestions and guidance.

I heartily express my sincere thanks to Dr. R.D. Tripathi, Scientist, National Botanical Research Institute, Lucknow for valuable suggestions.

I am extremely thankful to Dr. U.N. Rai, Scientist, National Botanical Research Institute, Lucknow, for his guidance and suggestions.

I express my sincere and heartful thanks to Dr. K.P. Singh, Scientist, Industrial Toxicology Research Centre, Lucknow, for his fruitful suggestions and cooperation.

With pleasure, I record my sincere thanks to Dr. U.P. Dwivedi, Dr. M.K. Singh and Mr. N.K. Mishra for their keen interest and discussion from time to time.

I record my sincere thanks to all my fellow researchers Mr. A.K. Dwivedi, Mrs. Tripti Pandey and Mr. A.K. Pandey.

I cannot forget the untiring help, keen interest and moral support of my friend Mr. Ritu Srivastava during the research work.

The staff at the Feeder Balancing Dairy, Ramnagar, Varanasi (Ramnagar Dairy) are gratefully acknowledged for their technical assistance.

I thank Dr. T.B. Singh, I.M.P. Banaras Hindu University, Varanasi for valuable suggestions for statistical analysis.

I would like to thank Mr. D.K. Sharma and Mr. A.K. Rai for their help and cooperation.

I would also like to thank my parents and brothers, who have supported me all through this course without whom I would not have attained the position at which I am now.

Alka R. Upadhyay

PREFACE

Environmentalism in the 1980s was characterized by a conscious shift from enthusiastic rhetoric to development of alternative ways to solve environmental problems associated with local, regional, and global issues, such as industrial waste, acid precipitation, global warming, ozone depletion and human population. People and institutions began puffing more energy and resources into solving environmental problems. These approaches to the environment in the 1970s and 1980s saw many advances as well as many failures. I believe many failures resulted from a basic lack of understanding about the environment and how natural ecological systems function.

In recent years our understanding of many aspects of the environment has increased greatly. Degradation of our surface water and ground water resources is a serious problem, the effects of which may not be fully known for some time. There are a number of steps we can take to treat water and to minimize pollution. Nutrients released by human activity lead to water pollution. Two important nutrients that can cause eutrophication and pollution problems of water are phosphorus and nitrogen, both of which are released from a variety of sources.

The dairy industry of India has grown from an almost completely unorganised into a vastly complex organised industry of a large magnitude during the last forty years. India today ranks first among milk producing countries. Dairy industry is noted as a significant contributor to pollution the wastes are characterized by a relatively high organic concentration, high oxygen demand, nitrogen and phosphorus.

Since long, many chemical and mechanical methods were being employed to treat such polluted waters, that proved to be cost effective and time taking. Recently, biological treatment of water bodies is gaining attention of the environmentalists. This initiated me to undertake a detailed investigation of the dairy industry regarding processing, waste water generation, evaluation of conventional treatment plant (primary and secondary treatment), and to develop effective and economic biological methodology for waste water treatment. During present investigation efficiency of five aquatic macrophytes have been tested in individual as well as in different combinations besides the quality of raw and treated waste water.

The book has been divided into eight chapters. Chapter I deals with introductory part of the study. Chapter II covers the review of literature pertaining to the present work, followed by Chapter III which describes meteorological conditions, location of study site, dairy industry functioning, sources of waste water generation and conventional treatment method.

Chapter IV deals with various analytical methods and statistical analysis for various physico-chemical and biological parameters. Chapter V incorporates the results of different physico-chemical parameters of dairy waste water (raw waste water-influent and treated waste water-effluent).

Chapter VI describes the waste water treatment by aquatic macrophytes. Chapter VII includes general discussion of various physico chemical parameters, milk procurement and waste water generation and aquaculture study. Chapter VIII deals with summary, conclusions and recommendations.

The relevant references and appendices have been listed in the last.

Alka R. Upadhyay

CONTENTS

LIST OF TABLES

CHAPTER – I

INTRODUCTION

Waste Water Generation

Every community produces both liquid and solid wastes. The liquid portion wastewater is essentially the water supply of the community after it has been fouled by a variety of uses. From the stand point of sources of generation, wastewater may be defined as a combination of the liquid or water carried wastes removed from residences, institutions, and commercial and industrial establishments (Metcalf and Eddy, 1991).

Pollution of surface waters occurs when too much of an undesirable or harmful substance flows into a body of water, exceeding the natural ability of that water body to remove the undesirable material, dilute it to a harmless form.

Types of Water Pollutants

Water pollutants are categorised as emitted from point or non-point sources. Point sources are distinct and confined, such as pipes from industrial or municipal sites that empty into streams or rivers. Non-point sources, such as runoff, are diffused and intermittent and are influenced by factors such as land use, climate, and geology. Common urban non-point sources include urban runoff from streets or fields. Such runoff contains all sorts of pollutants from heavy metals to chemicals and sediments (Botkin and Keller, 1995). All segments of our society (urban, rural, industrial, agricultural, and military) may contribute to the problem of water pollution.

Almost all industries discharge water containing wastes from some stage of their manufacturing process, but industrial wastes are not same in every case. Every industry produces its own characteristic process wastes, as a result of many operations such as washing operation, washing of raw materials, formation of intermediates and final products, washing of product containers etc. The wash waters usually contain a small quantity of the process material, which imparts the characteristics to the process wastes.

Adverse Effects of Waste Water

Waste water produces biochemical oxygen demand and diseases. Phosphorus and nitrogen from agricultural and urban land use, fertilizers and waste water from sewage treatment causes artificial eutrophication. Nitrates in ground water and surface waters can cause pollution and damage to ecosystems and people. Heavy metals released from agricultural, urban and industrial use of mercury, lead, selenium, cadmium etc. causes significant ecosystem damage and human health problems. Pollution caused due to sediments released from runoff from construction sites, agricultural runoff and natural erosion, reduces water quality and results in loss of soil resources.

Nutrients that stimulate oxygen consumption by bacteria and other decomposers in a river or lake, for instance, may be lethal to fish, but will stimulate a flourishing community of decomposers. Although the types, sources, and effects of water pollutants are often interrelated, it is convenient to divide them into following major categories:

1. Health problems – Infectious agents, organic chemicals, inorganic chemicals and radioactive materials.
2. Ecosystem disruption – Sediment, plant nutrients, oxygen demanding wastes and thermal.

Water clarity is affected by the abundance of plankton organisms and is a useful measure of water quality and water pollution. Eutrophication, an increase in nutrient levels and biological productivity, is a normal part of successional changes in most lakes. Tributary streams bring in sediments and nutrients that stimulate plant growth. The high biological productivity of eutrophic system is often expressed as "blooms" of algae or thick growth of aquatic plants and high levels of sediment accumulation. Eutrophication also occurs in marine ecosystems, especially in nearshore waters and partially enclosed bays or esturies. Blooms of minute organisms called dinoflagellates produce toxic red tides that kill fish.

Water pollution problems in surface waters are often both highly visible and a direct threat to environmental quality. Like most developed countries, the United States and Canada have made encouraging progress in protecting and restoring water quality in rivers and lakes over the past fifty years. In 1948, only about one third of Americans were served by municipal sewage systems and most of those systems discharged sewage without any treatment or with only primary treatment (the bigger lumps of waste are removed). Most people depended on cesspools and septic systems to dispose of domestic wastes. The 1972 Clean Water Act established a National Pollution Discharge Elimination System (NPDES), which requires an easily revoked permit for any industry, municipality or other entity dumping wastes in surface waters (Botkin and Keller, 1995).

Since the passage of the Clean Water Act in 1972, the United States has spent more than $ 100 billion in public funds and much more in private investments to control these "conventional" point sources of water pollution. About one third of the public expenditures have been used to build or upgrade thousand of municipal sewage treatment plants. As a result by 1990, 70 per cent of the U.S. population was served by municipal sewage systems. Four-fifths of these plants had secondary or tertiary treatment and major city was discharging raw sewage into a river or lake except as overflow during heavy rain storms.

Passage of the 1970 Water Act in Canada has produced comparable results. Seventy per cent of all Canadians in towns over 1,000 population are now served by some form of municipal sewage treatment.

Japan, Australia and most of Western Europe also have improved surface water quality in recent years. Sewage treatment in the wealthier countries of Europe generally equals or surpasses that in the United States. Sweden for instance serves 98 per cent of its population with at least secondary sewage treatment (compared with 70 per cent in the United States), and the other 2 per cent have primary treatment. Denmark and Germany have municipal sewage treatment of 90 per cent and 84 per cent of their populations respectively. The poorer countries have much less to spend on sanitation. Spain serves only 18 per cent of its population with ever primary sewage treatment. In Ireland, it is only 11 per cent and in Greece, less than 1 per cent of the people have even primary treatment. Most of the sewage, both domestic and industrial, is dumped directly into the ocean.

This lack of pollution control is reflected in inland water quality as well. In Poland, 95 per cent of all surface water is unfit to drink. The Vistula River, which winds through the country`s most heavily industrialized region, was so badly polluted in 1978 that only 432 of its 1,068 km were suitable even for industrial use. It was reported to be "utterly devoid of life" (Botkin and Keller, 1995).

According to C.P.C.B., 90% of the water supplied in India to the towns and cities is polluted, out of which only 1.6% gets treated. Of the nation's 3119 towns and cities, only 8 boast of complete sewage collection and treatment facilities. According to the Ministry of Environment and Forests (MEF), most rivers in India are polluted, mainly because of direct inflow of untreated sewage resulting in unacceptable levels in them of biological oxygen demand (BOD) and suspended solids (SS). For example, the Damodar river is the most polluted river, carrying discharges from 43 major industries and scares of major units. In Yamuna (Delhi), 19,000 cubic metres of water containing DDT derivatives from agriculture are dumped daily. 68 industrial units discharge 1000 cu.m. of untreated waste water in Ganga every day. 25% of the total pollution of Ganga is from industrial waste from Calcutta and Kanpur. Industrial and municipal waste discharges have made segments of Ganga unfit for bathing. The Ganga carries industrial and untreated sewage from 114 cities, each having a population of more than 50,000. 75% of the pollution of Ganga is from untreated waste (sewage), 88% of which comes from 25 class I cities (Sharma, 2000).

Impact of water pollution on aquatic life

Most of the rivers and fresh water streams in India are badly polluted by industrial wastes or effluents. Contamination of the industrial wastes which come in the water is most dangerous. The sewage of big cities is often drained into rivers. The sewage promotes the growth of phytoplanktons. This excessive growth depletes the oxygen of water. This reduction of oxygen and the presence of poisonous wastes affect the fish population. Besides this, rivers, lakes and ponds are also used directly by people for bathing or washing. This contaminates the water with germs of various diseases like cholera, dysentery, hepatitis, typhoid, gastroenteritis, snail fever, enteric fever and malaria.

The effluents produce physical, chemical and biological changes in water. Some pollutants produce only temporary effects on water whereas others have long standing effects. Liquid effluents of industries containing a variety of poisonous chemicals are discharged into the bodies of waters. They not only change the pH of water but also adversely affect the aquatic plant and animal life and sometimes cause large scale killing of fish and other aquatic animals.

Due to addition of domestic waste water concentration of phosphate, nitrate and other nutrients increases in the water bodies. This enrichment of the nutrients is called eutrophication. Thus through the process of organic matter decomposition enriched in terms of nutrients which help in highly increasing productivity. With the addition of nutrient, luxuriant growth of algae in water has been reported by numerous workers. Shifting of algal flora i.e. blue green algae to predominate have also been reported. These start forming algal blooms, floating scums or blankets of algae. Blooms of algae are generally not utilized by zooplanktons. The algal blooms compete with other aquatic plants for light for photosynthesis. Thus oxygen level of water is depleted. Pollutants like suspended particles, etc. cause turbidity which ultimately reduce the penetration of light, and result in decreasing or in the elimination of aquatic plants. Dissolved chemicals in high concentrations become toxic and reduces productivity of aquatic system. Many of our lakes, notably the Dal lake, Naini lake, Ambazari lake, Powai lake, Deshalser Talao, Rewalsar lake and Ranmal lake are becoming darkened, smelly and choked with excessive growth of algae. The ponds which were generally used for domestic purposes, fish culture, growing *Trapa bispinosa* and other edible crops, they are often covered by certain aquatic weeds, which poses treat to their multipurpose use.

Rivers, the important source of drinking and irrigation water, are extensively used to discharge untreated industrial effluents and city sewage. Due to this reason all the important Indian rivers such as Brahmaputra, Ganga, Cauvery, Godavari, Sabarmati, Mahanadi, Son, Narmada and Subarnrekha are being polluted continuously (Tripathi, 1986). Yamuna has been highly polluted in Delhi upto Okhla due to discharge of city and industrial wastes through Najafgarh drain which is an important source of pollution carrying effluents from 216 industrial units (Tripathi et al., 1984). River Chambal is continuously being polluted due to direct discharge of industrial wastes containing urea, ammonia, lead, mercury, cadmium and other toxic chemicals (Olaniya et al., 1976). River Son is heavily polluted due to direct waste discharge of paper mill and jute mill at district Shahdol, Madhya Pradesh. About 80 % of the Indian population is directly or indirectly dependent on the river waters. Further, 80 % of the infectious diseases are water borne and 50 % of the deaths among the children is due to diarrhoeal diseases (Manja, 1986).

A substantial portion of the total fresh water available in the hydrological cycle is needed to sustain natural aquatic ecosystem – marshes, rivers, coastal wetlands, and the millions of species that they shelter (Convention on Wetlands, 1998). Healthy natural ecosystem are indispensable regulators of water quality and quantity.

Treatment

Over the past one hundred years, sanitary engineers have developed ingenious and effective municipal waste water treatment systems to protect human health, ecosystem stability and water quality (Tables 1.1 and 1.2). Primary treatment is the first step in municipal waste treatment. It physically separates large solids from the waste stream. A moving screen that filters out plastic packets, bottles and similar smaller items. Brief residence in a grit tank allows sand and gravel to settle. The waste stream is then pumped into the primary sedimentation tank about half the suspended, organic solids settle to the bottom as sludge. Secondary treatment consists of degradation of the remaining suspended solids. The effluent from primary treatment is pumped into a trickling filter bed, an aeration tank or a sewage lagoon. Aeration tank digestion is also called the activated sludge process. Effluent from primary treatment is pumped into the tank and mixed with a bacteria rich slurry. Air or pure oxygen pumped through the mixture encourages bacterial growth and decomposition of the organic material. The remainder would be valuable fertilizer if it were not contaminated by metals, toxic chemicals and pathogenic organisms. Tertiary treatment removes plant nutrients, especially nitrates and phosphates, from the secondary effluent. Although waste water is usually free of pathogens and organic materials after secondary treatment, it still contains high level of inorganic nutrients, such as nitrates, phosphates, iron, potassium and calcium. When discharged into surface waters, these nutrients stimulate algal blooms and eutrophication. To preserve water quality these nutrients must also be removed.

Table 1.1 : Chemical treatment

Treatment Method	Type of Waste	Mode of Operation	Degree of Treatment	Remarks
Ion Exchange	Plating, Nuclear	Continuous filtration with resin regeneration	Demineralized water recovery, product recovery	May require neutralization and solids removal from
Reducing & Precipitation	Plating	Batch or continuous treatment	Complete removal of chromium and heavy metals	One day's capacity for batch treatment 3 hrs. retention for continuous treatment, sludge disposal or dewatering required
Coagulation	Paper board, refinery, rubber, paint	Batch or continuous treatment	Complete removal of suspended and colloidal matter	Flocculation and setting tank or sludge blanket unit, pH control required

Wetland disposal works well in some areas. Marshes and bogs have a great capacity for absorbing nutrients and other pollutants. In some cases artificial marshes have been constructed as a cheap but effective way to control water pollution. The biomass they produce can be used to generate methane, a valuable clean burning fuel. They also serve as a home for Wildlife and a pleasant addition to the urban landscape.

Table 1.2 : Conventional waste treatment

Treatment Method	Lagoons	Aerated lagoons	Activated sludge	Tricking filter	Spray irrigation
Mode of Operation	Intermittent or continuous discharge, facultative or anaerobic	Completely mixed Continuous basin	Completely mixed or plug flow; sludge recycle	Intermittent or continuous application; may employ effluent recycle	Intermittent application of waste
Degree of Treatment	Intermediate	High in summer, less in winter	> 90 % removal of organics	Intermediate or high depending on loading	Complete; water percolation into ground water and runoff to stream
Land Requirements	Earth dry, 10-60 days retention	Earth basin, 6-12 ft. deep; 8-16 acres/mgd	Earth or concrete basin, 10-15 ft. deep 7,500-350,000 ft3/mgd	225-1,440 ft3/mgd	40-3 00 gpm/acre
Equipment		Pier-mounted or floating surface aerators	Diffused or mechanical aerators; clarifier for sludge separation and recycle	Rock filters 3-8 ft. deep; plastic packing 20-40 ft deep	Aluminium irrigation and spray nozzles, movable for relocation
Remarks		Solids separation in lagoon; periodic dewatering and sludge removal	Excess sludge dewatering and disposed of		Solids separation required; salt content in waste limited

Low Cost Waste Treatment

The municipal sewage systems used in developed countries are often too expensive to build and operate in the developing world where low-cost, low-tech alternatives for treating wastes are needed. One option is *effluent sewerage,* a hybrid between traditional septic tank and a full sewer system. Another alternative is to use natural or artificial wetlands to dispose of wastes, Arcata, California for instance needed an extensive sewer plant upgrade. By transforming a 65 hectare (160 acre) garbage dump into a series of ponds and marshes that serve as a simple, low-cost waste treatment facility, the city saved millions of dollars and improved the environment simultaneously. Sewage is piped to holding ponds where solids settle out and are digested by bacteria and fungi. Effluents flow through marshes where it is filtered and cleaned by aquatic plants. The marsh is a haven for wildlife and has become a prized recreation area for the city. Eventually, the purified water flows into the bay where marine life flourishes.

Similar wetland waste treatment systems are now operating in many developing countries. Effluent from these operation can be used to raise fish for human consumption, if care is taken to first destroy pathogens. Usually 20 to 30 days of exposure to sun, air, and aquatic plants is enough to make the water safe. These systems make an important contribution to human food supplies. A 2,500 hectare (6000 acre) waste fed aquaculture facility in Calcutta, for example, supplies about 7000 metric tons of fish annually to local markets.

Industrial waste treatment may be required for a variety of reasons depending on the location of the industrial plant, the regulations governing the discharge of effluents, and the availability and economics of process water. The major reasons can be summarised as follows :

1. Reduction of existing pollution – In many parts of the country the pollution loads discharged to receiving streams are excessive. As a result partial or complete treatment may be required for all industries to a particular stream.

2. Plant expansion or production increase – Many states have set maximum limits on the quantity of BOD (Biochemical Oxygen

Demand) or suspended solids, which can be discharged to a stream. Even when a waste treatment plant exists increases in production resulting in higher pollution loads require additional or expanded waste treatment. In some cases permits to expand production facilities must be accompanied by plans for increased waste treatment capacity.

3. New plant location – It is becoming increasingly common to consider waste treatment requirements as one of the factors governing selection of a plant site, where no present regulations governing effluent requirements exists, a stream survey should be run to determine the allowable pollutional loading and in turn the waste treatment requirements.
4. Water reuse – Industrial expansion on major and minor water courses, with accompanying increases in demand for process and cooling water, has sometimes resulted in production limitations during periods of low stream flow. Although waste treatment may not be necessary to abate stream pollution, treatment of some or all the waste may be essential to provide needed process water (Eckenfelder W., 1966).

Use of Aquatic Plants for Waste Water Treatment

The capacity of vascular aquatic plants to assimilate nutrients from polluted waters has been recognized for several years (Rogers and Davis, 1972; Steward, 1970; Boyd, 1976). Vascular aquatic macrophytes such as water hyacinth (*Eichhornia crassipes* [Mart] solm.), duckweed (*Lemna minor*) and cattails (*Typha* spp.) cultured in ponds and reservoirs offer potential alternatives for treating sewage and industrial effluents (Boyd, 1969; Wooten and Dodd, 1976; Wolverton and Mc Donald, 1979) and agricultural effluents (Reddy et al., 1982). Nutrient removal efficiency of a system containing plants will depend on the type of aquatic plant, growth rates of plant, nutrient composition of water and physico-chemical environment in the water.

Aquatic macrophytes which have rapid growth rates and absorb large quantities of nutrients might provide a practical and economic method for tertiary treatment of effluent. The aquatic macrophytes use solar radiation and thus have a low energy requirement as compared to other methods of tertiary treatment.

Aquatic plants utilize the nutrients and produce large amounts of biomass which can be used for some beneficial purposes. Aquatic macrophytes systems can be effectively used to reduce pollutant levels in water bodies (Boyd, 1969; Lakshman, 1979; Stowell et at., 1981; Reddy et aL, 1982) and the biomass used for production of gaseous fuels (Shiralipour and Smith, 1984) feed (Bagnall et al., 1974) fiber (Nolan and Kirmse, 1974) and compost and organic soil amendments (Parra and Hortenstein, 1974).

Studies reported by several researchers (Clock, 1968; Scarsbrook and Davis, 1971; Cornwell et al., 1977) calculate the nutrient removal rates, based on the changes in concentrations at the inflow and outflow of a pond or reservoir. Although these calculations provide information on the nutrient removal efficiency from waste water, they provide very little understanding on the rate of N and P removal in these systems. Many research workers have discussed the nitrogen and phosphorous removal capacity of different aquatic plants (Wolverten et al., 1979; Aoyama et al., 1986; Busk et al., 1989). Culley and Epps suggested that duckweeds might be desirable plant for removing nutrients from sewage effluent. In general duckweed species are more tolerant of cool weather than water hyacinth. Also, duckweed would be easier to handle than water hyacinth because it does not form the dense entangled mat that is characteristic of water hyacinth. However most of these studies are restricted to one or very few plants, there is lack of comparative data among different plants grown under the same environmental conditions.

Dairy Industry and Pollution

Milk has served as a food for mankind even before the beginning of recorded history. Before urbanization each family depended on its own animals for milk. Later on dairy firms were developed close to cities. Feeder balancing dairies were installed nearby metropolitan cities to collect milk from rural milk-shed areas for supply it to metropolitan dairy to fulfil the requirement of urban areas. The available surplus milk is conserved as SMP (skim milk powder) and white butter which is used in reconstitution in lean months for supply as liquid milk.

The dairy industry of India has grown from an almost completely unorganised into a vastly complex organised industry of a large magnitude during the last forty years. India today ranks first among milk producing countries. However, every thesis generates antithesis. Dairy industry is noted as a significant contributor to pollution. It is true that by adopting recent methods of mechanised handling, the quantity of milk lost to the drain is reduced. However, hygienic requirements in the industry result in huge amount of waste water generation, mostly from dilution of milk and milk products. The wastes are characterised by a relatively high organic concentration, high initial oxygen demand, nitrate and phosphate (Marshall and Harper, 1984). Dairy wastes exert a high rate of deoxygenation, the initial oxygen demand is high, about fifty per cent of the total oxygen demand being exerted in twenty four hours. The rate of deoxygenation is more than twice that of domestic sewage (Strom, 1974). Plants with fully automated product processing have above average waste coefficient (Lytken, 1974). Dairy waste do not contain significant quantities of toxic compounds, although accidental spillage of cleaning and sanitizing chemicals may constitute a hazard. Excessive quantities of dairy wastes are detrimental to normal aquatic life because of oxygen depletion. The presence in the water of chironomid larvae and turbificid worms, which are tolerant of adverse conditions, is normally an indication of heavy pollution. If all the dissolved oxygen is removed, the lactose in the waste will be converted to organic acids and case in precipitated, forming black benthal deposits. Any suspended solids in the waste water will settle and smoother the bed of a slow moving stream. Nutrients such as phosphorus and nitrogen may also have an effect, causing the growth of a certain organisms such as algae and fungi. A particular problem caused by the discharge of milk wastes is that even quite large dilutions of milk are turbid so that relatively small quantities can cause cloudiness in natural waterways.

Ramnagar dairy (Dugdh Utpadak Sahkari Sangh Ltd., Ramnagar, Varanasi) is a feeder balancing dairy. Dairy procures about 1 lakh Litre milk daily from district Varanasi and other nearby districts and marketing of milk and milk products like ghee, paneer, butter and milk cake is done. Dairy is located on the eastern bank of river Ganga and drains treated effluent in the upstream. Dairy waste is organic in nature. It is rich in nutrients, nitrate and phosphate, which is source of enhanced algal growth in natural waterways (Marshall and Harper, 1984).

In dairy industry, primary and secondary treatment methods are quite common in the treatment of dairy waste water, as they are efficient and dependable. The dairy effluent is predominantly organic in nature and due to its biodegradable constituents, it is amenable to conventional treatment. It is probably due to this reason that most of the existing dairies have treatment plants based on activated sludge process. This type of treatment method is not effective in filtering the nutrients from the dairy waste water. This waste water needs further polishing to remove the nutrients which can be effectively done through the use of aquatic macrophytes.

The purpose of present investigation was to quantify the dairy waste water and examine physico-chemical properties of raw and treated waste water. Present study also aims to develop an effective and economic biological methodology for the waste water treatment.

Aims and Objectives

The proposed research work aims to analyse physico-chemical properties of raw and treated waste water and to evolve effective and economic biological treatment method for the treatment of dairy waste water using aquatic macrophytes.

- General survey of the study site.
- Quantification of physico-chemical properties of untreated and treated dairy waste water.
- Aquaculture experiments to evaluate the efficiency of aquatic macrophytes for best possible removal of nutrients from the dairy waste water.
- Statistical analysis to check the significance of variance of the observed data and correlate between some important parameters.
- To recommend effective and economic biological methodology for waste water treatment.

CHAPTER – II
REVIEW OF LITERATURE

Introduction

Environment is an aggregate of all the external factors and conditions which influences the life. The study of environment is the pavement of all the reactions that will reflect the problems to be faced in future. In the present day the technology known is exploiting the natural aquatic resources and thus, damaging the environment at an alarming rate. Today, the most urgent need is to examine the cause, extent and preventive measure of pollution of the natural aquatic resources, which are commonly used for multifarious purposes by a huge population.

Water Chemistry and Biological Property in Relation to Pollution

Physico-chemical and biological properties of fresh water bodies ih relation to pollution was observed by number of workers (Hill, 1979; Nielson, 1975; Reddy, 1981; Reddy and Graetz, 1981; Bruce et al., 1982; Burden et al., 1987; Pearl, 1987; Grobbelaur, 1989; John Stone, 1975; Felfoldy, 1972; Roush, 1985; Pavoni et al., 1990; Pritchard, 1985; Engler and Patrick, 1974; Duarte and Kaift 1990).

Among the many environmental aspects which influences the distribution of aquatic macrophytes, water chemistry has long been recognised as on important factor that limits the occurrence of many species (Hellquist, 1980; Kadono, 1982b; DeBusk and Dierberg, 1989).

Acute problem of eutrophication in aquatic bodies, diverted the researchers towards the relationship prevailing between the nutrients and algal production. The IBP programme emphasized on the theme of biological basis of productivity and human welfare, as a result Golterman et al., 1969 contributed the methods of physico-chemical analysis to the limnology. Some other contributions on this aspect were given by Reddy, 1981; Rai and Hill, 1984; Bombowna, 1985; Ame et al., 1988 and Vighi and Chiaudhan, 1985. Effect of season on nitrogen and phosphate contents present in pond water have been observed by Yoshimuro, 1932. Moorhed et al., 1988, studied the changes in nitrogen forms in a water hyacinth based water treatment systems. The effect of physico-chemical characteristics on the behaviour of nitrogen and phosphorus has been studied by Mikkelson et al., 1978; Reddy and Graetz, 1981; Bode et al., 1987 and for major ions and conductivity by Kingsford et al., 1973.

The studies pertaining to heavy metal pollution in industrial effluents and fresh water bodies have been studied by Hasan et al, 1986; Mathis and Kavern, 1975; Mortimer, 1985; Elfilne, 1987; Gerber et al., 1987; Cossa and Noel, 1987. Aquatic macrophytes have also been studied as pollution indicator by Cairns and Niederlhner, 1987; Mortimer, 1985, Wahlquist, 1972 worked on water quality changes by production of *Eichhornia crassipes*. Hepher, 1958 studied dynamics of phosphorus added to lentic system.

Ryther and Dunstan, 1971 studied eutrophication in coastal marine environment. Dierberg and Brezonik, 1981 demonstrated nitrogen fixation associated with decaying leaves of pond cypress in natural and a sewage-enriched cypress dome, Nitrogen and other nutrients are lost from floating plant mats in particulate detritus and through leaching (Gaudet, 1977).

Waste Water Analysis

Standard methods for examination of water and waste water (APHA, AWWA, and Water Pollution Control Federation, 1995) were used for present research work.

Hach, 1969 worked on water and waste water analysis. Murphy and Riley, 1962 derived modified single solution method for phosphate determination in natural waters.

Determination of total Kjeldahl nitrogen-N by flow injection analysis was studied by Tecator, 1992. Water analysis have been done for nutrients by Hill, 1979; Burden et al., 1987 and Pearl and Bowles, 1987 for heavy metals by Mathis and Kevern, 1975; Cossa and Noel, 1987 and Elfline, 1987, for trace metals, Timperley, 1978 and Abu Hilal, 1987. Certain other noteworthy contributions were of Rai and Hill, 1981a, 1982; McDonnel, 1982; Gerber et al., 1987 and Crumpton et al., 1987.

Removal of Nutrients by Aquatic Macrophytes

Aquatic macrophytes play a significant role in regulating the eutrophication status of fresh water bodies. The aquatic plants utilize the nutrients present in polluted water and produce large amount of biomass which can be used for some beneficial purposes. The concept of using aquatic plants for treating waste water is gaining attention of new researchers as Paterniti et al., 1986; Ingemarsson, 1986; Ruschel, 1987; Budd and Kerson, 1987; Ice and Couch, 1987 and Pip, 1987.

Vascular aquatic plants for mineral nutrient removal from polluted water (Boyd, 1969), Vascular aquatic macrophytes such as water hyacinths (*Eichhornia crassipes* [Mart] Saims), duckweed (*Lemna minor*), and cattails (*Typha* sp.) cultured in pond reservoirs, offer potential alternative for treating sewage and industrial effluents (Boyd, 1969; Wooten and Dodd, 1976; Wolverton and McDonald, 1979), and agricultural effluents (Reddy et al; 1982), Culley and Epps (1973) suggested that duckweeds (Lemnaceae) might be a desirable plant for removing nutreints from sewage effluents. The capacity of vascular plants to assimilate nutrients from polluted waters has been recognized for several years (Rogers and Davis, 1972, Steward, 1970; Boyd, 1976). Studies reported by several researchers (Clock, 1968; Scarsbrook and Davis, 1971; Cornwell et al., 1977) calculated the nutrient removal rates, based on the changes in concentration at the inflow and outflow of a pond reservoir. Aquatic macrophytes can be effectively used to reduce pollutant levels in water bodies (Stowell et al., 1981; Reddy et al., 1982) and the biomass used for production of gaseous fuels (Shiralipour and Smith, 1984) feed (Bagnall et at., 1974), fiber (Nolan and Kirmse, 1974), and Compost and organic soil amendment (Parra and Hortenstein, 1974). Several studies have discussed the potential of aquatic plants for reducing N and P levels in waste water (Sutton and Ornes, 1975; Dunigan et al., 1975; Reddy, 1983). Several studies have dealt with nutrients removal from sewage by harvesting planktonic algae, but rather elaborate and expensive techniques are required to harvest phytoplankton (Golueke, C.G. 1964). Several authors (Boyd 1968; Burgess, 1965; Mackenthum, 1964, 1965) have hinted at the possibilities of using higher aquatic plants for nutrients removal. Water hyacinth cover on water chemistry (Charles McVea and Boyd, 1975) alters the physico-chemical characteristics pH, dissolved oxygen, alkalinity, and temperature of the water. Change in these characteristics can influence the behaviour of Nitrogen and phosphorus in the water (Bouldin et al., 1974; Mikkelsen et al., 1978; Reddy and Gractz, 1981). Low dissolved oxygen in hyacinth ponds can favor the NO_3-N loss through denitrification (Reddy et al., 1980).

Aquatic macrophytes also promote N removal by lowering the water and thereby enhancing sedimentation of particulate matter, after containing organic nitrogen and $NH4^+$ (Howard-Williams, 1985).

The aquatic plants studies, involving waste treatment include the water hyacinth (Wolverton et al., 1976; Wolverton and McDonald, 1976; Dinges, 1978; Cornwell et al., 1977; Schuize, 1966), duckweed (Culley and Epps, 1973; Sutton and Ornes, 1977; Bartsch, 1961; Seidel, 1976) and submersed plants (McNabb, 1976). Most researchers recognize the water hyacinth as the most prolific of these aquatic plants.

Water hyacinths were found to be less efficient in removing phosphorus than nitrogen (Steward, 1970; Dunigan et at, 1975; Boyd, 1976). Cornwell et al., 1977 studied nutrient removal by water hyacinth.

Water hyacinth is the most promising species for nutrient removal due to its prodigious growth rate and free floating nature (Boyd, 1969). It has been reported that one acre of water hyacinths can potentially assimilate from domestic sewage the nitrogen generated by 595 persons and the phosphorus generated by 180 people producing 67 tons of diy matter annually (Steward, 1970).

Aowal and Singh, 1981 discussed treating dairy waste water by *Eichhornia*. DeBusk et al., 1989 studied performance of water hyacinth based secondary treatment system. Obeid and Chadwick, 1964 worked on factors affecting the growth of two aquatic weed species.

Dunigan et al, 1975 used water hyacinths to remove nutrients from eutrophic waters, Peters et al., 1980 characterized and compared five N_2-fixing *Azolla-Anabaena* associations. Madsen and Adams, 1988 studied nutrient dynamics of a submersed macrophyte community.

Wolverton et al., 1976 studied pollution removal properties of aquatic plants and their other benefits. Wolverton and McDonald, 1976 used *Eichhornia crassipes* for upgrading sewage lagoons.

Nitrogen is lost rapidly from decomposing aerial water hyacinth tissues (Ogwada et al., 1984 and DeBusk, 1982). Boyd, 1969 studied mineral nutrient absorption by weed plants. *Lemna* sp. was used for nutrient removal study by Harvey and Fox (1973). Duckweed plants are able to grow successfully on waste water to convert the degradable pollutants directly into useful materials, such as protein rich fodder. Studies and field monitoring indicate that DWs grow relatively well on sewage water (Culley et al., 1981). On livestock wastes (Myers, 1977), under laboratory conditions on several pollutants such as NH_4^+, PO_4^{3-}, on anaerobic water (Wolverton and McDonald, 1981) and even on water bodies polluted with detergents (Agami et al., 1976).

Aquatic Plants and Biomass Production

Under central Florida conditions, Yount and Grossman, 1970 measured a growth rate of 5-29 g dry wt m^{-2} day^{-1}. Tucker and DeBusk, 1981 observed growth rates of 19.3 and 14.2 g dry wt m^{-2} day^{-1} (December-June) respectively, for water hyacinth and water lettuce cultured in nutrient enriched water. Reddy and DeBusk, 1984 studied the growth characteristics of aquatic macrophytes cultured in nutrient enriched water.

Algae are considered among the most prolific of the fresh water plants. Maximum obtainable growth rates for these plants vary greatly. Oswald, 1976 reports yields of 35.2 to 70.4 t/ha/yr for algae harvested from enriched sewage lagoons. However, McGarry and Tongkasome, 1971 report that yields of 157 t/ha/yr are obtainable when algae are grown year-round. One hectare of water hyacinths grown in an enriched environment in a warm climate for seven months of the year can be used to produce approximately 58,400 m^3 (2,290,000 scf) of biogas containing 35,100 m^3 (1,370,000 scf) methane (Wolverton and McDonald, 1979).

Knipling et al., 1970 studied nutrient content and yield potential of *Eichhornia crassipes*. Sculthorpe, 1967; Penfound and Earle, 1948 and Holm et al. 1969 worked on the biology of aquatic macrophytes. McBay, 1961 worked on biology of *Tilapia aurea*. Edmondson, 1974 worked on fresh water biology. Bradshaw et al., 1964 investigated mineral nutrition of several grass species.

Vollenweider, 1969 studied methods of measuring primary productivity. Penfound, 1956 worked on primary production of vascular aquatic plants. Westlake, 1963 did comparisons of plant productivity.

Various scientists worked on germination and propagation of water hyacinth (Haigh, 1940; Manson and Manson, 1958; Barton and Hatchkiss, 1951). Obeid, 1962 investigated into the mineral nutrition of some weed species. Minshall and Scartb, 1952 studied effect on water morphology of water hyacinth grown in acid media.

Jackson, 1980 worked on marine biomass production. Arnott, 1966 studied calcification in plants. Raschke, 1968 worked on waste reclamation and primary production in tertiary sewage stabilization pond. Raschke, 1970 conducted study on algal periodicity and waste treatment in pond ecosystem.

Chloride may be significant to growth of some plants in trace amounts (Levitt, 1969). Lawrence and Weldon, 1965 worked on identification of aquatic weeds.

Vetter, 1972 indicated cattle find *Eichhornia crassipes* palatable under certain conditions. Pirie, 1960 discussed water hyacinth, and Bates and Hentges, 1974 studied favourable and unfavourable effect on water hyacinth.

McGarry and Tongkasome, 1971 worked on water reclamation. Alich and Inman, 1976 worked on energy from agriculture. The giant algae or Kelp *(Macrocystis)* that has been considered a prime candidate for bioconversion produces an average yield of 14.5 t organic matter/ha/yr (North, 1971).

Center and Spencer, 1981 reported water hyacinth's competitive success is its flexible canopy configuration, which enables stands of this plant to maximize energy (solar radiation) capture while minimizing non-photosynthetic (supportive) plant structure. Natural stands of water hyacinths, for example, typically attain a standing crop of 2 kg dry wt m^{-2} (Penfound and Earle, 1948). Water hyacinth roots, which comprised the bulk of the detritus deposited in the tanks, decompose more slowly than aerial tissues (DeBusk and Dierberg, 1984).

A rapidly growing stands of water hyacinth assimilates Ca 1.3 g Nm^{-2} day^{-1} (Reddy and DeBusk, 1985). Tucker and DeBusk, 1981 reported that tissue N concentrations of water hyacinth cultured under constant nutrient availability were high in the winter, but low in the summer due to the dilution effect of rapid dry matter production. Water hyacinth can maintain a dense standing crop, as well as its flexible morphology, by utilizing tissue N accumulated during periods of high N availability (DeBusk and Dierberg, 1989).

Peckol et al., 1994 studied nutrient uptake capacities and tissue constituents of the macroalgae. Liaw and MacCrimmon, 1978 assessed changes in biomass of periphyton. Lock, 1993 studied attached microbial communities in fresh water.

Rogers and Davis, 1972 after measuring nutrient uptake by *Eichhornia crassipes* in growth chamber experiments, concluded that absorption by 1 ha. of water hyacinths would exceed 2,500 kg of nitrogen and 700 kg of phosphorus per year if maximum growth could be sustained. Many aquatic plants, including water hyacinth, have a high protein content and plants removed to reduce nutrient enrichment of natural waters could probably be used as animal feed (boyd, 1968, Boyd, 1969). Wahlquist, 1972 demonstrated that nitrogen and phosphorus were the nutrients limiting growth of water hyacinths in ponds. Previous growth rate studies by Dymond, 1949 and Penfound and Earle, 1949 were used by Westlake, 1963 in estimating the annual productivity of the water hyacinth to be 11-13 t/ha. dry weight. A later study by Wooten and Dodd, 1976 found a production of 30 t of organic matter ha^{-1} in only 105 days.

Tiedje, 1988 studied ecology of denitrification and dissimilatory nitrate reduction to ammonium. Svensson, 1993 did denitrification measurements in the Kaliby ponds. Sorensen et al., 1988 studied denitrification in stream epilithon. Sorensen and Revsbech, 1990 worked on denitrification in stream bioflim and sediment. Planas et al., 1996 worked on relationship of nutrient of *Cladophora glamerata* in two lake basins of different trophic status.

Buresh and Patrick, 1978 worked on nitrate reduction to ammonium in anaerobic soil. Stensel et al., 1973 studied biological kinetics of suspended growth denitrification. Yamane, 1957 conducted study on nitrate reduction and denitrification in flooded soils. Raveh and Avnimelech, 1973 worked on minimizing nitrate seepage. Luijn et al., 1996 compared denitrification rates in lake sediments obtained by N_2 flux method. Howard-Williams, 1985 studied cycling and retention of nutrients in wetlands. Brye, 1970 summarized nutrient concentrations and entrapment of reservoirs.

Bouldin et al., 1974 conducted study on losses of inorganic nitrogen from aquatic systems. Bremner, 1965 studied inorganic forms of nitrogen. Reddy et al., 1980 conducted study on nitrate reduction in soil-water system.

Nutrients Removal Through Constructed Wetlands

Constructed wetlands have received a great deal of interest in recent years for the treatment of various kinds of waste waters (Hammer, 1989; Moshiri, 1993; Kadlec, 1995). According to Brix, 1994 constructed wetlands have potential for secondary treatment of municipal and some industrial waste waters as well as polishing of tertiary treated waters. While many constructed wetlands are effective in consistently reducing biochemical oxygen demand and total suspended solids (Gearheart and Higley, 1993), nitrogen removals are more variable (Hammer and Knight, 1993). Flooding of soil creates anoxic (oxygen-deficient) condition under these conditions nitrate-nitrogen (N03-N) is used as an electron acceptor during microbial respiration, a process widely known as denitrification (Reddy et al., 1980).

Several studies showed that water hyacinth and duckweed will remove phosphorus from nutrients-enriched water (Burgess, J.E. 1965, Culley and Epps. 1973, and Scarsbrook and Davis, 1971). Increased pH as a result of increased rate of photosynthesis over respiration can enhance NH_3 volatilization losses (Bouldin et at., 1974 and Mikkelsen et at., 1978). Reddy and Graetz, 1981 observed rapid nitrification in a water system containing dissolved CO^2.

Lai and Lam, 1997 studied major pathways for Nitrogen removal in waste water stabilization ponds. Function of algae as an oxygen producing sources for aerobic bacterial decomposition

(Bartsch, 1961), the results of this study concur with previous findings that algae can play a direct role in nitrogen removal in waste stabilization ponds (Tom et at., 1975; Ferrara and Avci, 1982; Santos and Oliveira, 1987).

Several studies have shown light induced O_2 inhibition of denitrification in biofilms (Nielsen et al, 1990; Sorensen and Revsbech, 1990 and Dalsgaard and Revsbech, 1992). At abundances of submersed vegetation, epiphytic denitrification would probably have more impact on N removal (Eriksson and Weisner, 1997). This indicates, that denitrification in epiphytic communities in dense submersed vegetation can be of similar importance as denitrification in the sediments.

A preliminary report on the potential use of *E. crassipes* to remove nutrients in an anaerobic lagoon effluent system was made by Miner et al., 1971. *Eichhornia crassipes* seems to offer great potential value in reducing inorganic nutrients in waste waters because it reproduces vegetatively at a very rapid rate (Penfound and Earle, 1948) and chemical evaluations indicate a relatively high rate of utilization of essential elements (Boyd, 1969; Boyd and Vickers, 1971). Rogers and Davis, 1972 designed laboratory experiments to determine absorption of nitrogen and phosphorus by water hyacinths.

Previously biochemical oxygen demand, suspended solids and faccal coliform bacterial count have commonly been used as criteria for assessing the efficiency of sewage treatment facilities, while nutrient parameters have largely been overlooked (Toms et al., 1975 and Hussainy, 1979). More attention should be paid to the control of nitrogen in water bodies receiving points source discharges where nitrogen is often limiting for algal growth (Gakstatter et al., 1978). Although waste stabilization ponds have been widely used over the world, and proved to be an economical way of sewage treatment (Gloyna, 1971), waste stabilization ponds often do not have special design configurations for nutrient removal.

Eichhornia crassipes is apparently exerting upon another aquatic, water lettuce (*Pistia stratiotes* L.). Gay, 1958 noted that *P. stratiotes*, abundant in the White Nile before the advent of water hyacinth, had been virtually eliminated some regions and an 'antagonism' between the two species has also been reported

(C.S.A. 1958). This point is of interest in connection with observation made upon the extent to which *Eichhornia crassipes* modifies the substrate in which it grows (Bishai, 1960 and Chadwick, 1961). *Eichhornia crassipes* and *Pistia stratiotes* differ considerably in their pH requirements for optimum growth (Chadwick and Obeid, 1966).

Bouldin et al., 1974 reported maximum NO_3 loss will occur in shallow bodies of water where the sediments have a high biological activity and hence relatively thin oxidized zone. Ryther and Dunstan, 1971 have reported that inorganic nitrogen is likely a factor limiting aquatic plant growth in the coastal marine region of New York city despite large inputs of nitrogen from sewage and other sources. Howeler and Bouldin, 1971 derived a model for describing losses by denitrification. Losses of NH_3 from ammonium fertilizers applied on the surface of agricultural soil has been well documented for a variety of experiments (Gasser, 1964). Under natural conditions, NO_3^- formed in the water would diffuse into the underlying sediment and undergo denitrification (Engler and Patrick, 1974).

Azolla cultured in N free medium accumulated about 35 g N kg^{-1} of plant tissue, primarily due to symbiotic N_2 fixation through an *Azolla-Anabaena* symbiotic relationship (Peters et al., 1980).

Studies in America and Europe have shown that the nutrient removal efficiency by stabilization ponds was higher in summer than in winter (Toms et al., 1975; Pano and Middlebrooks, 1982; Santos and Oliveira, 1987). Nitrate loss through denitrification has been widely studied by several research workers (Bremner and Shaw, 1958a; Yamane, 1957; Stanford et al., 1975a and Burford and Bremner, 1975) for various soil types and under varying environmental conditions. In agricultural soils planted to crops, denitrification is undesirable because of NO_3-N loss to gaseous end products. The beneficial use of this process was recognized the waste water treatment process, where NO_3-N removal is the primary objective (Dawson and Murphy, 1972; Raveh and Avnimelech, 1973 and Stensel et al., 1973). The rate of NO_3-N removal from flood water is influenced by depth of flood water pH, temperature, energy source, and flood wafer concentration. The influence of some of these factors on denitrification in soil systems has been studied by several research worker Bremner and Shaw, 1958b; Nommik, 1956; Stanford et al., 1975b).

Oxygen diffusion into the flood water can also interfere with the NO_3-N reduction process, because facultative aerobes prefer O_2 over NO_3-N as an electron acceptor (Alexander, 1977). Studies reported by several researchers (Clock, 1968; Scarsbrook and Davis, 1971; Cornwell et al., 1977) calculate the nutrient removal rates, based on the changes in concentrations at the inflow and outflow of a pond or reservoir.

Cooke (1994) studied nutrient transformation and showed that approximately 34 per cent of influent NO_3^- to a waste water treatment wetland was transformed to NH_4^+ via the dissimilatory pathway. Nutrient accumulation in waste water treatment facility by periphyton was studied by Davis, Hoffmann and Cook, (1990). Gersberg et al., 1986 performed study on waste water treatment in artificial wetland by using aquatic plants. Various scientists used constructed wetlands for effluent treatment (Thakral et al., 1994; Moshiri, 1993; Hammer and Knight, 1993; Hammer, 1989; Gearheart and Higley, 1993 and Brix, 1994). Chevalier and Noue, 1985 studied nutrient removal from waster water by using algae. Shelef et al., 1980 researched on production of algal mass as an integral part of reclamation system. Huang et al., 2000 studied nitrogen removal in constructed wetlands.

Tarn and Wong, 1989 worked on nutrient removal by *Chlorella pyrenoidosa* and *Scenedesmus sp.* Rodrignes and Oliveira, 1987 treated waste water from the tomato concentrate industry in high rate algal ponds.

Diel Variation in Fresh Water

Diel variations in the physico-chemical properties of aquatic bodies are the result of interactions among numerous environmental factors (e.g. morphometry, climate, daily weather, and local topography). Environmental interactions determine die! variation in physico-chemical parameters of water bodies (Philip, 1927). Photosynthesis is the principal cause of the high concentration and saturation values of dissolved oxygen. High pH levels and large diel pH fluctuations in lakes are often associated with photosynthesis (Philip, 1927). pH increases downstream in reservoirs (Goldman and Wetzel, 1963; Neel, 1963). These changes have been linked to the increasing homogeneity of physico-chemical conditions downstream in reservoirs (Goldman and Wetzel, 1963).

Dye et al., 1980 worked on diel variations of physico-chemical parameters in fresh water. Changes in dissolved CO_2, HCO_3, and $CO\frac{2}{3}$ levels of water are controlled by the changes in pH of the water, and equilibrium reactions among these three components (Reddy, 1981). Ganf, 1974 worked on diurnal mixing and distribution of phytoplankton. Dunn, 1967 studied diurnal fluctuations of physico-chemical conditions in freshwater pond.

Ovon, 1936 studied waste water treatment and renovation by different duckweed species. Tchobanoglous et al., 1989 worked on evolution and performance of pilot-scale aquatic waste waster treatment using *Eichhornia crassipes*. Weber and Tchobanoglous, 1985 studied nitrification in water hyacinth treatment systems. Tchobanoglous, 1987 worked on engineering considerations for waste water treatment using aquatic plants.

Mitchell, 1973 worked on the growth and management of *Eichhornia crassipes* and *Salvinia* spp. in their environment and in alien situations and Dale and Giliespie, 1976 studied influence of floating vascular plants on the diurnal fluctuations of temperature near the water surface. Ultsch and Anthony, 1973 worked on the role of gaseous exchange in the ecology of *Eichhornia crassipes*. Chadwick and Obeid, 1963 reported on response of variation nutrients of some weed species. Various researchers worked on effect of water hyacinth on the ecology (Jepson, 1933; Gay and Berry, 1959; Anon, 1951; Bishai, 1960; Gay, 1958). Reddy et al., 1981 studied aquatic ecosystems as a means of agricultural drainage water treatment efficiency of nutrient removal by reservoirs and flooded fields.

Waste Water Treatment

Amending coarse sands, such as Bassendean sand, with a clay material (red mud) was found to improve both phosphorus and nitrogen removal from primaxy and secondary effluent (Ho et al., 1992). The phenomenon of enhanced biological phosphorus removal by activated sludge in excess of normal metabolic requirements was referred to as luxury uptake (Bargman et al, 1971; Levin and Shapiro, 1965; Connell and Vacker, 1967; Wells, 1969; Barchardt and Azad, 1968). Numerous studies have been conducted on various aspects of the mechanisms of that control this biological process (MeLaren et al., 1976; Hong et al., 1982; Spector 1977; Tracy and Flammino, 1987; Fuchs and Chen, 1975; Davelaar et al., 1978; Nicholls and Osborn, 1979; Berber and Winter, 1984; Aniw et al., 1988; Claete and Steyn, 1988).

Like most biological reactions, the organisms in the enhanced biological phosphorus removal process favour a near neutral pH. The results of an extensive laboratory study showed that the maximum phosphorus uptake rate is obtained in a pH range of 6.8 to 7.4 (Krichten et al., 1985). However, experience in full scale operations indicates that the effects of pH on the EBPR are more severe (Hong and Andersen, 1993). Untreated domestic waste water typically contains 20-50 mg/I of total nitrogen (Metcalf and Eddy, 1991). Typically, assimilation removes 20-30% of the total influent nitrogen (Van Haandel et al., 1981).

Biological denitrification reduces nitrate (NO_3^-) to nitrogen gas (N_2), nitrous oxide (N_2O) or nitric oxide (NO). This nitrogen removal process is the one most widely used in municipal waste water treatment (Water Poll. Con. Fed. 1983). Low DO is the most critical condition since denitrification is simply Several modifications of the aerobic pathway used for BOD oxidation (U.S. EPA, 1975).

Bonhomme et al., 1990 studied nitrogen removal in activated sludge. Emori et al., 1994 immobilized nitrifiers with polymer materials and introduced them to the aeration tank of an activated sludge process in the form of pellets. Oswald et al., 1957 studied use of algae in waste reclamation. King and Smith, 1947 investigated effects of water hyacinth on fish.

Calorimetriç studies of biodegradation were conducted by Fortier et al., 1980 in waste treatment plant. Mikkelsen et al., 1978 studied ammonia volatilization losses from flooded rice soils. Stratton, 1968 performed on ammonia nitrogen losses from streams. Gaudet, 1977 studied accumulation and loss of nutrient by papyrus in tropical swamps. Clock, 1968 studied nitrogen and phosphorus removal from secondary sewage treatment effluent.

Allsopp, 1960 and Parham, 1947 worked on weed control studies. Wolverton, 1980 studied higher plants for recycling human waste. Wolverton, 1982 conducted study on hybrid waste water treatment by use of *Phragmites cammunis* and microorganisms. Wolverton and McDonald, 1981 worked on treatment of organic chemical waste. Wunderlich, 1967 worked on control of aquatic vegetation. Removal costs of $ 25 to $ 35 per ton for surface and submersed weeds, respectively, have been contrasted to chemical control costs of $ 8 to $ 12 per ton (Wunderlich, 1968).

Algae have a voracious appetite for phosphorus and nitrogen and are extremely efficient at purifying wastewater (Lind, 1998). Lind, 1998 studied phosphorus inactivation in waste water treatment. Waste stabilization ponds are simple treatment system ideally suited for tropical developing countries because of their low cost, simplicity in operation and high efficiency even at fluctuating hydraulic loads (Arceivala, 1986). Ionized orthophosphate is believed to be the most important form for algal nutrition (Wetzel, 1983). All other forms of phosphorus (condensed and organic phosphorus) eventually are hydrolysed into the ortho form (Sawyer and McCarty, 1978). Einstein and Hunter, 1967 studied hydrolysis of condensed phosphate. Nalewajko and Lean, 1980 worked on phosphorus in the physiological ecology.

Nitrogen removal from waste water has been discussed by various scientists (Leakovic et al., 2000; Schuch et al., 2000; Lorenzen, 1999). Choung and Jean, 2000 worked on phosphorus removal in waste water.

Removal of Heavy Metals

The plant remove metals by surface adsorption and/or absorption and incorporate them into their own system or them in a bound form. Effluent ameliorated by these plants thus causes less damage to the aquatic environment. Of the various organisms assessed for their suitability in metal removal, only immobilized algae and bacterial capsules have gained prominence (Lavoic and De La Noue, 1983; Beveridge, 1989). Not only do the aquatic macrophytes take up metals from water and sediments and release them subsequently on aging and decomposition (Mayes et al., 1977, McIntosh et al., 1978), they also serve as food for animals in which metals are concentrated (Newman and McIntosh, 1983).

Although much literature exists on the uptake of metals by various macrophytes (Merchyulenene and Nyanishkene, 1976; Pip,

1990; Manny et al., 1991) the emphasis has only been on the cycling of metals from sediments to macrophytes. Macrophytes seem to offer a better alternative due to various reasons, they can accumulate high concentrations of toxic metals (Charpentier et al., 1987). Bryophytes have been found to accumulate high levels of heavy metals; of which a major fraction (70-90%) lies adsorbed on the cell surface (Beckett and Brown 1984; Brown and Beckett, 1985; Wells and Brown, 1987, 1990). Researches have demonstrated the ability of aquatic macrophytes in reducing the level of toxic metals in polluted waters Wolverton and McDonald, 1979; Wolverton 1981; Chigbo et al., 1982; Zirchky and Reed, 1988; Palbo and Weaks, 1990; Everard and Denny, 1985; Outridge, 1992). Works have demonstrated that aquatic macrophytes can be used to partially strip trace metals in the waste waters (Seidel, 1971; Wolverton et al., 1975; Brik and Schierup, 1989; Crowder, 1991; Guilizzoni, 1991). Various scientists worked on accumulation of heavy metals by water hyacinth (Sutton and Blackburn, 1971; Tukunga et at., 1976; Yeoh, 1979). Piccardi and Clauser, 1983 worked on *Iris pseudocasis* on absorption of metal. Sela et at., 1989 studied accumulation of heavy metals by water fern.

Chigbo et al., 1982 worked on uptake of heavy metals by water hyacinth. Sinicorpe et al., 1992 studied metal removal by wetland mesocosms. Staves and Knaus, 1985 worked on metal removal by using different species of duckweeds. Guilizzoni et al., 1989 worked on ecology and heavy metal concentration in aquatic macrophytes. Yashizaki and Tomida, 2000 worked on process of metal removal from sludge.

Dairy Waste Water

Dairy effluents essentially contain residues of milk, milk product intermediates, cleaning chemicals, suspended and dissolved inorganic solids from blow down/bleed water. A dairy centre that processes 80,000 m^3 of milk per year generates approximately 540 m^3 of waste water (Marshall, 1978). The strong environmental impact (Nemerow N.L., 1976) of this high volume of effluents, makes its purification of great interest. A treatment will be considered effective if the concentrations of the parameters that limit dumping are reduced below the legally permitted values; or if it enables good use to be made of their constituents.

Brown and Pico, 1980 characterized dairy waste and worked on treatment of dairy waste. Various scientists worked on dairy processing and waste water management (Carawan, Chambers and Zall, 1979; Carawan and Jones, 1977; Hall and Hedrick 1966; Fjaervoll, 1970; Tuszynski, 1978). Strom (1974) worked on parameters expressing the pollution. Svoboda et al., 1966 worked on purification of dairy waste water. Svoboda, 1974 worked on waste stabilization ponds. Various scientists studied treatment of dairy waste water (Wheatland, 1974; Vandamme and Waes, 1980; Trebler and Harding, 1955; Hemmings, 1980; Doedens, 1974a; Seyfried, 1974; Boyle and Polkowski, 1973; An Foras Taluntais, 1974). Adamse, 1966, 1968 worked on bacteriological studies on dairy waste. Barnett et al., 1982 worked on characteristics of dairy effluent. Various scientists worked on disposal of dairy waste water (Galpin, 1981; McDowell and Thomas, 1961; Magnusson, 1974; Schraufnagel, 1957; Riddle and Chandler, 1974; Parkin and Marshall, 1976; Watson et al., 1977; Arbuckle, 1970). Various scientists worked on development in dairy effluent treatment (Harper and Chambers, 1978; Horton and Trebler, 1953; Synnott et al., 1978; Cooper 1974). Marshall, 1978 worked on characteristics of dairy effluent. Carta et al., 1999 worked on aerobic purification of dairy effluent. Harold, 1974 studied pollution control in the dairy industry. Hanne, 1978 worked on methods of analysis for dairy waste water. Antonie and Welch, 1969 worked on dairy waste treatment. Bosset and Blanc, 1978 determined colour of milk and milk products. Burton, 1984 studied changes in milk at high temperatures.

Keeney and Bassette, 1959 studied browning reaction in milk. Castellfort, 1978 worked on solar powered milk pasteurizer. Various scientists worked on problems of odours (Baker, 1961; Symons, 1956; Rosen et al., 1962; Geldard, 1953). Rosen et al., 1963 worked on relationship of odour to specific contaminants. Various scientists worked on odour control (Middlebrooks, 1965; Kuehner, 1964; Silvey et al., 1965).

Various scientists worked on residues of organochlorine pesticides in milk (Abdrabo et al., 1989; Shaker et al., 1988; Jodrai et al., 1995).

Mechsner and Wuhrmann, 1974 worked on ecological consideration and biological treatment for dairy wastes. Porges et al., 1960 worked on dairy waste treatment. Various scientists worked on reduction in dairy waste (Royal, 1978; Trebler and Harding, 1955; Baltjes 1978). Buxton et al., 1977 and Kearney, 1973 worked on regulations of water quality of dairy processing plants. Neeteson, 2000 worked on N and P management on dairy farms.

Effect of Pollution on Aquatic Plants

The waste water produce physical, chemical and biological changes in water. The sewage promotes the growth of phytoplanktons. Waste water from industries, municipalities, tanning and slaughtering plants discharged into the rivers, streams, lakes, etc. may be a potential source of infective bacteria and other microbes which cause diseases in man and other animals.

Mangi et al., 1978 studied effect of metal on some aquatic plants. Outridge, 1992 compared Cd toxicity tests with plants. Cain et al., 1980 worked on toxicity and bioaccumulation of heavy metal in green alga. Forstner and Wittman, 1983 worked on metal pollution in aquatic environment. Charpentier et al., 1987 worked on toxicity and bioaccumulation of heavy metal in aquatic macrophytes in experimental cultures. Crist et al., 1981 studied nature of bonding between ions and algal cell walls. Jana, 1988 worked on changes in physiological and biochemical parameters due to accumulation of pollutants aquatic species. Sela et al., 1989 studied effect of pollutants on the water fern. Wang, 1986 researched on toxicity tests of pollutants by using aquatic macrophytes. Wong et al., 1980 studied metal toxicity to phytoplankton. Kozlowoski, 1986 measured the effects of pollution on plants. Experimental acidification effect on periphytic algae was studied by Parent et al., 1985. Lazarek, 1983 researched on structural and functional aspects of algae in the acidified lake. Hendrey, 1976 studied effects of pH on algae. Hendry and Vertucci, 1980 worked on plant communities in acidic lake, Almer et al., 1978 worked on pollution and the aquatic ecosystem. Baily and Stokes, 1984 evaluated algae as biomonitors of metal accumulation.

Crisman et al., 1980 studied biotic response due to pollution. Findlay, 1984 worked on effect of pH on phytoplankton Havens, 1984 researched on responses of plankton to acidification. Effect of nutrient enrichment on micro-algae was studied by Hillebrand and Sommer, 2000. Ault et al., 2000 worked on influence of nutrients in fresh water. Keddy et al., 2000 worked on effect of low and high nutrients on plants. Blanchard and Lereb, 2000 worked on effect of agricultural runoff. Ferrier Pages et al., 2000 worked on effect of nutrient enrichment.

Work Done in India

Numerous aspects of ecological investigations have been the subject of study by a large number of workers in India. Some investigations have also been made by a few workers on waste water treatment. Potential of water hyacinth for waste water treatment was studied by various scientists (Kumar and Garde, 1989; Kumar and Garde, 1990; Kumar, 1986; Lakshman, 1979; Shukla and Tripathi, 1989; Haider, 1984; Prakasham et al., 1998). Removal of fluoride from water was investigated using mainly organic waste materials and activated carbon (Seethapathy Rao, 1964; Sastry, 1975; Meenakshi et al., 1991; Mutukumaran et al., 1995; Sanjay Kumar, 1995).

Various scientists worked on impact of pollution on aquatic plants (Tripathi et al., 1990; Singh and Bhargava, 1985; Srivastava and Sahai, 1976). Rai et al., 1994 worked on waste water treatment by using aquatic plants. Various scientists worked on removal of heavy metals by use of aquatic plants (Tripathi and Chandra 1991; Selvapathy and Sreedhar, 1991; Munshi et al., 1989; Sinha et al., 1993; Rai and Chandra, 1989; Garg and Chandra, 1990; Chandra and Sinha, 1992; Rai and Chandra 1992; Sarkar and Jana, 1986; Jam et al., 1990; Joshi et al., 1982; Bayed, Semi and Hussaini, 1985; Rai and Mallick 1992; Jana, 1988).

Heavy metal pollution study in fresh water were conducted by Mathur et al., 1987. Goel and Trivedi, 1984 studied effect of waste water disposal to fresh water. Parija, 1934a and 1934b did physiological investigations on *Eichhornia crassipes* and on reappearance of water hyacinth. Bose, 1945 worked on problem of water hyacinth. Haque and Sharma, 1986 studied role of water hyacinth in pollution abatement.

Jhingran et al., 1969 worked on methodology on fresh water fisheries. Rai and Datta Munshi, 1977 observed diurnal fluctuation in physico-chemical properties of tropical swamps. Rao and Mathur, 1975 studied eradication of water hyacinth. Khan and Siddiqui, 1972 worked on waste water treatment. Rao et al., 1985 worked on treatment and disposal of waste water. Murthy et al., 1984 studied treatment and disposal of waste water. Tripathi et al., 1998 conducted study on nutrient contents in fresh water aquatic macrophytes. Kaul et al., 1980 worked on mineral removal by aquatic macrophytes. Kaul, 1981 studied role of macrophytes in freshwater ecosystems. Mukherjee, 1982 worked on treatment of waste water by aquaculture. Gajghate and Reddy, 1989 worked on relationship of biochemical oxygen demand and chemical oxygen demand. Sammaiah and Sastry, 1991 worked on dairy waste water treatment. Tripathi et al., 2000 worked on physico-chemical characterization of dairy waste water. Upadhyay and Tripathi, 2001 studied efficiency of aquatic macrophytes in purification of dairy waste water. Sayed, 1997 worked on pollution control in dairy plants. Khan et al., 1970 studied diurnal variations in freshwater pond. Sreenivasan et al., 1975 worked on diurnal and seasonal variations in freshwater pond. Raina et al., 1984 worked on water pollution. Various researchers worked on effect of pollution problem in rivers (Nandan, 1985; Ganapati and Chacko, 1951; Rajgopalan et al., 1970; Singh and Bhowmick, 1985; Rajgopalan et al., 1973).

CHAPTER – III

DAIRY INDUSTRY AND STUDY SITE

Milk Processing Industry in India: An Overview

The growth of milk processing industry in India gained momentum since the first "operation flood" was pressed for implementation in early seventies. The operation flood was aimed to capture for the public dairies a "commanding share" of the milk market in the four metropolitan cities of Mumbai, Delhi, Kolkata and Chennai. These dairies were in turn to be supplied with liquid milk from modern processing dairies to be located in rural areas with high milk production potential (called milk-sheds), and or gifted skimmed milk powder and milk fats for recombination. The rural dairies also helped balancing seasonal fluctuations in milk production through the separation of milk into fat and other solids in the surplus season and their storage for subsequent use in the lean season. It is reported that about 12 MLD milk is processed in Indian dairies. During milk processing about 42 MLD waste water is generated (N.D.D.B., 1993).

Milk Processing Industries : Classification

Experiences led to the deliberation that the dairies should be located where milk is generated, i.e. at the heart of the milk production centre. Such rural dairies should be linked up with an urban milk distribution system. Hence, the concept is of feeder/balancing dairies. City dairies are linked up with feeder/balancing dairies and are capable of recombination and raw milk processing. The difference of feeder and feeder balancing dairy lies on milk powder production. Feeder balancing dairies are product plant with powder production. Categorisation of industries on the basis of operational view point are:

- Liquid milk plant
- Feeder dairy plant (product plant without powder production)
- Feeder balancing dairy (product plant with powder production) &
- Chilling centers

Location of Milk Dairies

Most of the milk processing units are located in the States of Uttar Pradesh, Maharashtra, Gujarat, Tamil Nadu, Rajasthan, Madhya Pradesh and Punjab. In other states, the number of units is not so significant. In fact, milk procurement in Uttar Pradesh is high but processing units are not in tune with milk procurement. Karnataka and Tamil Nadu offer scope for more expansion of milk processing units. Milk processing units in Gujarat and Maharashtra are commensurate with respective milk procurement.

Cost of Treatment

Cost of treatment depends on two important factors. The first and the foremost is characterization of waste, and the design flow. The second one is the type of treatment. But these factors merge to achieve the standard laid down by regulatory authority.

Assumption is that waste water generation is 4 times the amount of milk processed. In practice it is observed that waste water generation lies between 3 to 5 times the milk processed. The optimal flow of 400 m^3/day is evolved on the basis of milk handling 1,00,000 Lpd of milk.

The equation established with flow constant at 400 m^3/day is, capital cost Rs. = 2,607 x B.O.D applied (Kg) + 1,12,4054. However, keeping B.O.D. constant at 1,200 mg/I the equation is :

Capital cost (Rs. in lakh) 5544 x Flow in m^3/day

The equations are on the basis of most economical alternative, that is Alternative III (Fig. 3.2).

Table 3.1 : Summarisation of Water Consumption (CPCB, 1993)

Cubic metre per day

Type of dairy	*Mother Dairy*		*Liquid Milk Plant*		*Feeder Dairy*		*Balancing Dairy*		*Chilling Plant*
Area of Consumption	**A**	**B**	**A**	**B**	**A**	**B**	**A**	**B**	
Process	200(14.3)	100(25)	530(37.86)	230(62.2)	1800(50)	450(50)	200(50)	300(30)	15(16.6)
Washing & Service	323(23.05)	140(35)	660(47.1)	60(16.2)	700(19.4)	100(11.1)	50(12.5)	.150(15)	50(55.5)
Cleaning in Process (CIP)	428(30.35)	45(11.25)	-	50(13.5)	-	-	100(25)	100(10)	-
Sanitary	250 (17.8)	45(11.25)	50(3.6)	20(5.4)	600(16.7)	200(22.2)	-	-	-
Utility	200(14.3)	70(17.5)	160(11.4)	10(2.7)	500(13.9)	150(16.7)	50(12.5)	450(45)	25(27.7)
Total	1401(100)	400(100)	1400(99.6)	370(99.8)	3600(100)	900(100)	400(100)	1000(100)	90(99.8)

Note : Figures in parenthesis indicate percent of total

A and B denote two different plants in each category

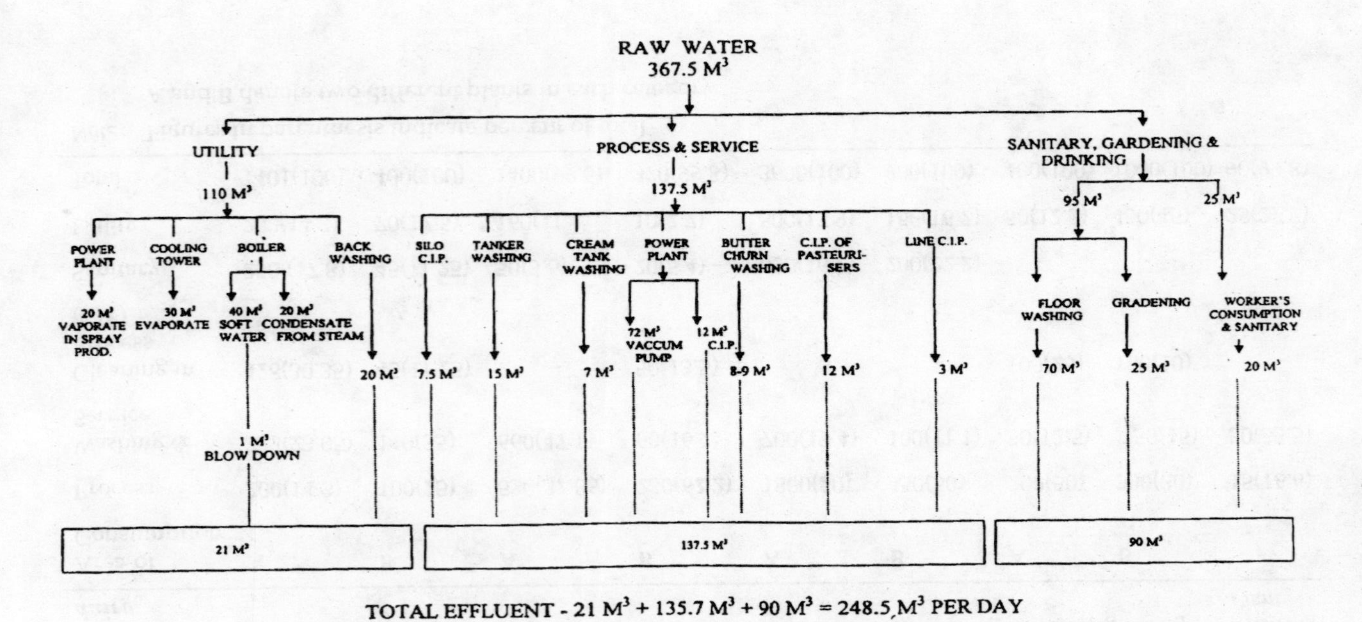

Fig. 3.1 : Mass Balance of Water Consumption and Waste Generation (Source : CPCB, 1993)

Table 3.2 : Product/Type Matrices of Dairy Industries (CPCB, 1993)

Product	Chilled raw milk received	Milk handled	Butter, white butter and butter oil	SMP & WMP	Ghee	Pasteurised Milk	Curd & Confec-tionaries	Paneer, cheese & casein
Type								
Mother dairy	++	+++	+	-	-	-	-	-
Liquid milk plant	+++	+++	+	+	++	+	-	+
Feeder dairies	+++	+++	++	-	++	-	+	++
Feeder balancing dairies	+++	+++	++	++	++	+	++	+++
Chilling centre	+++	-	-	-	-	-	-	-

Note : +++ - Maximum; ++ - Moderate; and + - Low

But the situation departs dramatically is the case of composite dairy which also produces butter, butter oil, cheese and other products would have a much higher organic load in the waste waters. It is observed that 95 % efficiency of BOD removal through option III are the cost effective solution.

Minimal National Standards (MINAS)

Generally two main aspects are taken into consideration for development of standards for waste water discharges. One relates to the adverse effects on health and environment and the other achievability of limits of pollutants by incorporation of appropriate pollution control measures.

The latter approach aims at use of the best available and economically feasible technology. Economically feasible technology assures that the cost of pollution control measures will remain within the affordability of the industrial units. Standards developed on these principles are techno-economic standards and these standards are uniform throughout the country.

An advantage of the technology based approach is that for a specific group of industries the extent of pollution control measures are alike. In addition, these standards serve to preserve the environmental quality in non-polluted areas. The disadvantage of this approach is that these standards may become unnecessary burden on the industry where the recipient environment does not demand such control measures. This is because these standards do not relate to the actual environmental situation of the specific site. In the case of dairy industry where the effluent is rich is organic load only secondary treatment does not serve the required quality to avoid the eutrophication of streams and rivers. Therefore, effluent polishing is necessary for the removal of nutrients, nitrogen and phosphorus.

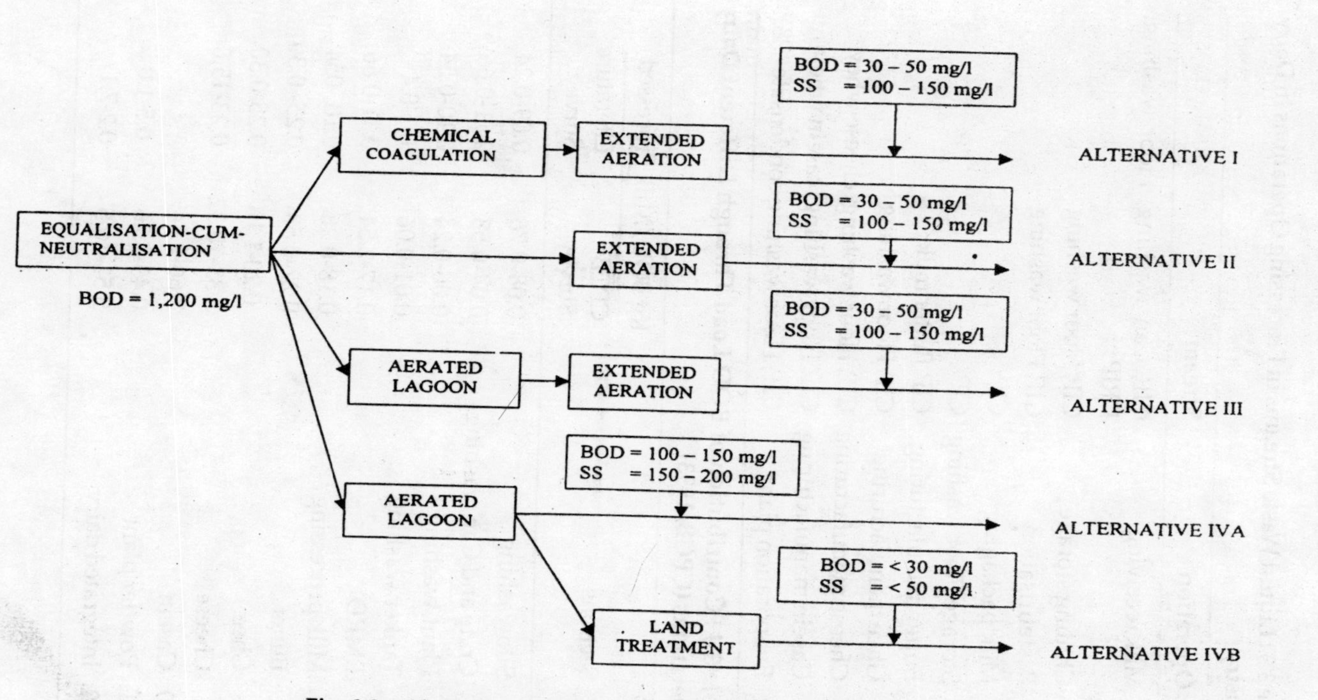

Fig. 3.2 : Alternative Schemes of Treatment for BOD Removal

Table 3.3 : Liquid Waste Streams in Processing Operations in Dairy Industry

S.N.	Operation	Stream
1.	Milk receiving	CIP, Can washing, Floor washing, DRIP
2.	Chilling storage	CIP, Floor washing
3.	Pasteurisation	CIP, Floor washing
4.	Milk packaging	CIP
5.	Storage Floor washing	CIP
6.	Butter manufacturing	CIP, Butter milk
7.	Ghee manufacturing	CIP, Floor washing
8.	Cheese manufacturing	CIP, Floor washing, cheese whey
9.	Casein manufacturing	CIP, Floor washing, casein whey
10.	Evaporation plant	CIP, Floor washing condensate

Table 3.4 : Contribution of BOD Load Through Different Dairy Operations (CPCB, 1993)

	Sections	Kg BOD/Milk Processed	
		CPCB Survey	Literature Survey
1.	Silowashing	0.08-0.76	0.09-0.24
2.	Crate and Can washing	0.04-0.53	0.11-0.66
3.	Plant washing	0.06-0.24	0.10-0.54
4.	Tanker washing	0.01-0.06	0.1-0.4
5.	RMRD	0.17-2.54	0.11-0.66
6.	Milk processing	0.18-0.28	0.10 -054
7.	Butter	0.27-0.53	0.25- 0.30
8.	Ghee	0.20-1.13	0.25-0.30
9.	Cheese	3.33-20.32	0.23-15.0
10.	Casein	2.66-2.74	
11.	Powder plant	0.43-1.26	0.5-1.0
12.	Integrated dairy	5.51-16.5	0.2-7.1

Study Site

For detailed investigation Ramnagar dairy was selected as study site. Details of the study site are as follows :

Geographical Position of Varanasi

Varanasi city lies at 25°18' N latitude and 83°1' E longitude in the middle of Gangetic plain at an average height of about 76.19 meter above the sea level, with even topography. This city covers an area of about 506181 hectare.

Meteorological Conditions of Varanasi

Climate

The climate of the region is tropical monsoonic type and is divisible into hot and dry summers (March-June), warm and moist rainy (July-October) and dry cold winter (November-February) with a few sporadic showers.

Temperature

The average mean ambient temperature ranges lowest in December (10.6°C in 1998 and 24.27°C in 1999) while average mean maximum temperature was highest during May (35.36-39.61°C) in 1998. The maximum average temperature gradually starts falling from June and reaches to the minimum during December when the winter is coldest, then it starts rising from January to May when the summer reaches to its peak. The rainy season months i.e. July to October normally found warm and wet. October was noted transition period from winter to summer. The day length was recorded longest in June (about 14 hrs.) and shortest in Dec. (about 10 hrs.).

Rainfall

The winter months were relatively dry with occasional rains. Maximum monthly rainfall 473.1 mm. was recorded in August 1998.

Humidity

Humidity varies widely in different seasons maximum being in the rainy and minimum in the summer season. The average maximum relative humidity during study period was in August, 1998 (93.1%) while the minimum was in April, 1998 (36.2%).

Location of Site

Ramnagar dairy (Feeder Balancing Dairy) (Dugdh Utpadak Sahkari Sangli Ltd., Ramnagar, Varanasi) is located on eastern bank of river Ganga, in Ramnagar Industrial Area, Ramnagar, Varanasi and drains treated effluent in the upstream.

Milk

Milk consists of 85-89% water and 11 to 15% total solids. The latter comprises solids-non-fat (SNF) and fat. Milk with a higher fat content also has higher solids-non-fat content with an increase in SNF of 0.4% for each 1% fat increase. The principal components of solids-non-fat are protein, lactose and minerals (ash). The fat content and other constituents of the milk vary with the species of animal and for the milk cow with improved breeding. Fat content and lactose of milk from buffalo which is the main source of Indian dairy is significantly high. The waste characteristics, therefore, vary with the type of milk fat which is a complex mixture of triglycerides (98-99%) phospholipids (1-2%) and traces of sterols, carotenoids and fat-soluble vitamins A, D, E and K. The vitamins exist as oil-in-water emulsion in small globules (2-5 μm) in the milk serum. The milk proteins comprises two major classes, the casein and whey proteins. The casein comprises 80% total protein. Lactose, a disaccharide molecule is a carbohydrate comprising one molecule of glucose and one of galactose. The major salts in the mineral component are those of calcium, sodium, potassium and magnesium, which occur as phosphates, chlorides, nitrates and caseinates. Sulphur, zinc, rubidium, silicon, bromine, aluminium, iron etc. are also present in trace quantities.

Table 3.5 : Average Composition of Milk (Handbook of Animal Husbandry, 1990)

	Water	*Fat*	*Protein*	*Lactose*	*Ash*
Buffalo	82.76	7.38	3.60	5.48	0.78
Cow	86.61	4.14	3.58	4.96	0.71

Milk Products

The basic function of dairy processing industry is the manufacfure of milk and milk products. Dairy products are such that it is difficult to classify and describe the processes involved. However, for determination of the significant sources of wastes from dairy product an understanding of dairy products, and their production processes are necessary.

Product Spectrum

Major dairy products are fluid milk, cream, butter and milk powder. In our country cultural diversification covers a wide spectrum of products like lassi, ice cream, ghee, milk cake, mishtidoi, flavoured milk, rasogolla and various other sweetmeats besides the products mentioned above. It is, therefore, expected that Indian dairies will not generate uniform waste due to variation of composition of milk and also the various types of products. Ramnagar dairy produces liquid milk, ghee, butter, lassi, flavoured milk and milk cake.

Milk procurement and waste water generation

Milk procurement was recorded higher in winter months followed by rainy and summer months. Milk was procured highest 83278.40 Litre in December (1999) (Fig. 3.5). Waste water generation was also recorded maximum 333113.60 Litre in December (1999). Lowest 15706.03 Litre milk procurement was recorded in August (1998) (Fig. 3.6). Waste water generation was also recorded lowest 50118.6 Litre in August (1998).

Processing

The processing operations for fluid milk or manufactured milk products include centrifugal sediment removal and cream separation, pasteurization and sterilisation, homogenisation and packing, handling and storing. Cooling is the most important unit operation in order to arrest enzyme activity and microbial growth. Cooling must be done to below 4.4°C.

Centrifugal devices include clarifiers, for removal of sediment, and extraneous particulates, and separators for removal of fat from milk. The fluid milk enters the separator at the centre under pressure with a positive displacement pump or centrifugal pump with flow control. The fat is separated and moves toward the centre of the bowl, while the skim milk passes to the outer space. Standardisation is the process of adjusting the ratio of butter fat and solids non-fat

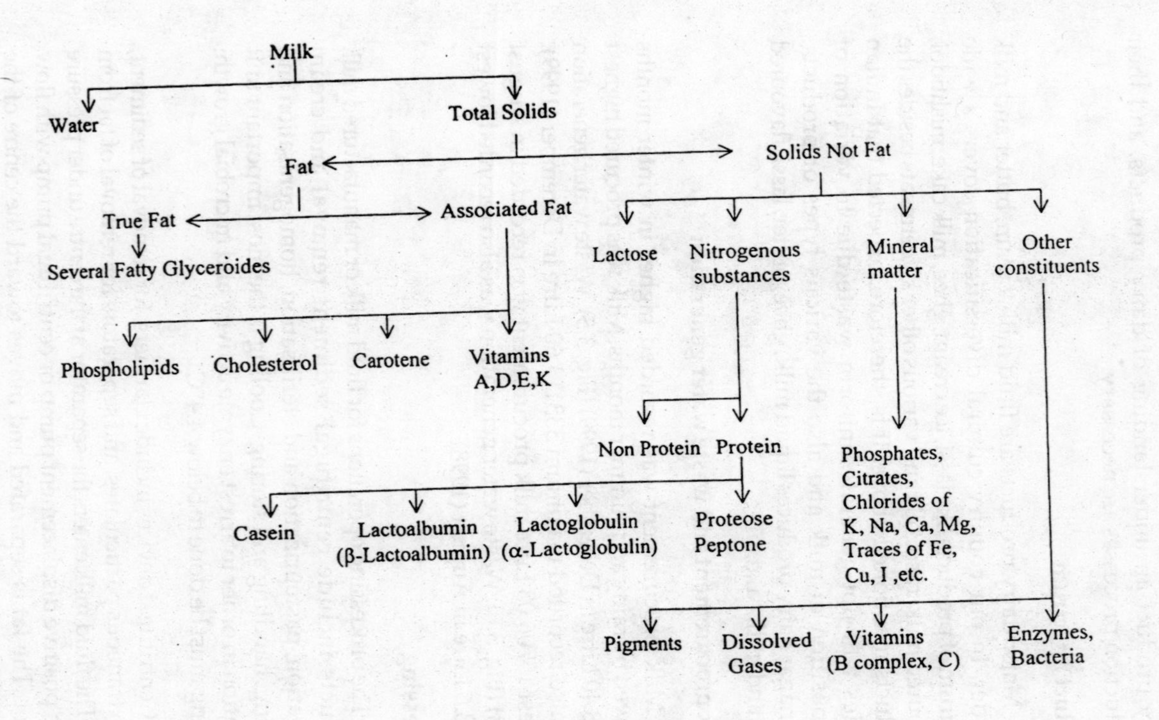

Fig. 3.3 : Milk Constituents

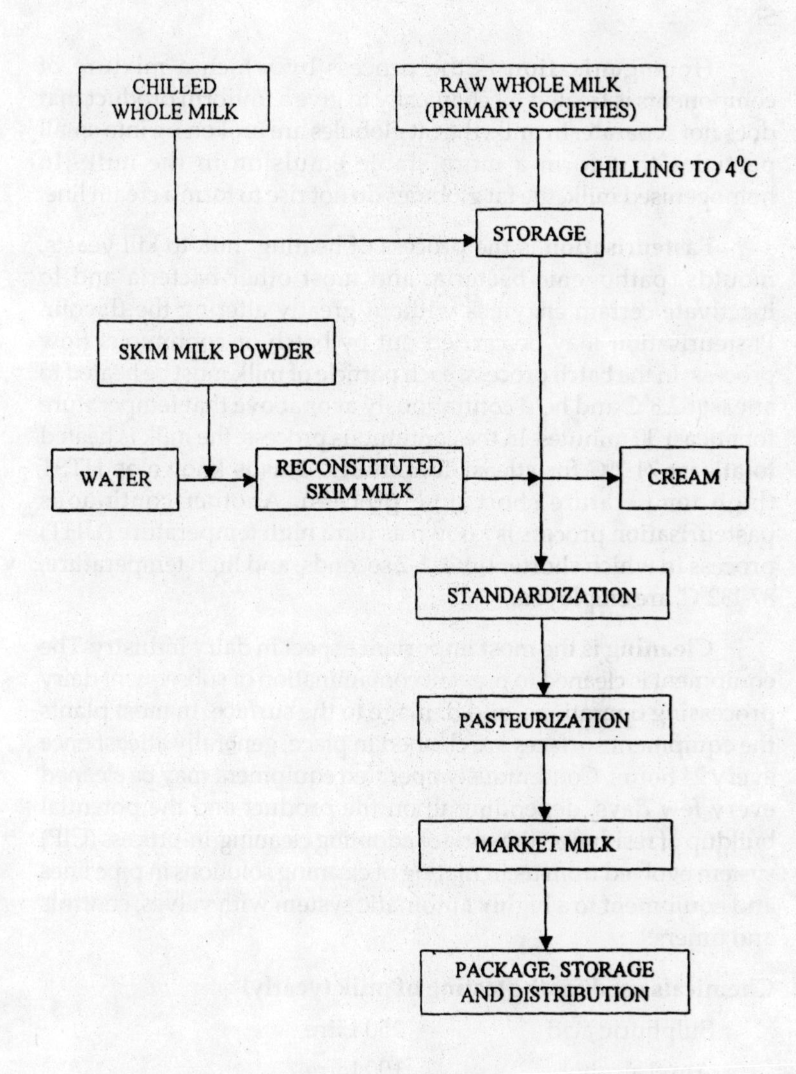

Fig. 3.4 : Process Flow Diagram for Market Milk Production

(SNF) to meet legal or industry standards. Adding cream of a high butter fat into low butter fat milk might result in a product with low SNF.

Homogenisation is the process by which a mixture of components is treated mechanically to give a uniform product that does not separate. In milk, the fat globules are broken up into small particles that form a more stable emulsion in the milk. In homogenised milk, the fat globuies do not rise to form a cream line.

Pasteurisation is the process of heating milk to kill yeasts, moulds, pathogenic bacteria, and most other bacteria and to inactivate certain enzymes without greatly altering the flavour. Pasteurisation may be carried out by batch or continuous flow process. In the batch process, each particle of milk must be heated to atleast 62.8ºC and held continuously at or above that temperature for atleast 30 minutes. In the continuous process, the milk is heated to atleast 71.7ºC for atleast 15 minutes. This is known as HTST (high temperature short time process). Another continuous pasteurisation process is known as ultra high temperature (UHT) process in which shorter time, 1-2 seconds, and high temperature, 87-132ºC, are employed.

Cleaning is the most important aspect in dairy industry. The equipment is cleaned to prevent contamination of subsequent dairy processing operations, and damage to the surface. In most plants the equipment surfaces are cleaned in place, generally atleast once every 24 hours. Continuously operated equipment may be cleaned every few days, depending upon the product and the potential buildup of residue on the surface adopting cleaning-in-process (CIP) system evolved from recirculating of cleaning solutions in pipe lines and equipment to a highly automatic system with valves, controls and timers.

Chemicals used for the testing of milk (yearly)

Sulphuric acid	250 Litre
Amyl alcohol	100 Litre
Ethyl alcohol	50 Litre
EDTA	10 kg

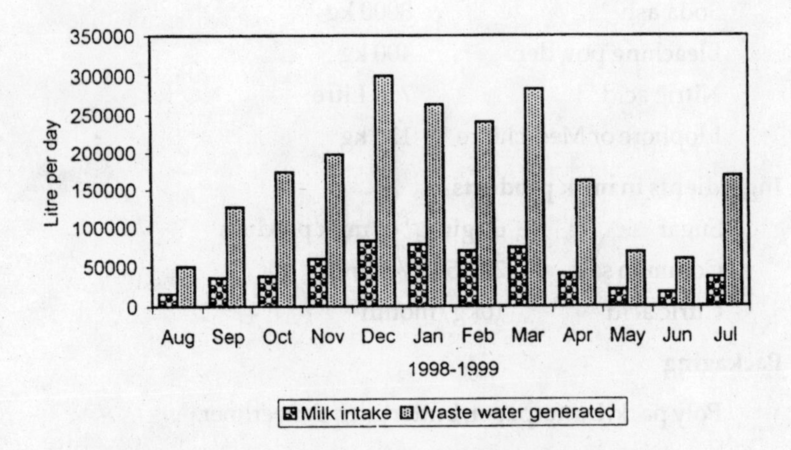

Fig. 3.5 : Monthly milk procurement and waste water generation (Ramnagar Dairy, Varanasi)

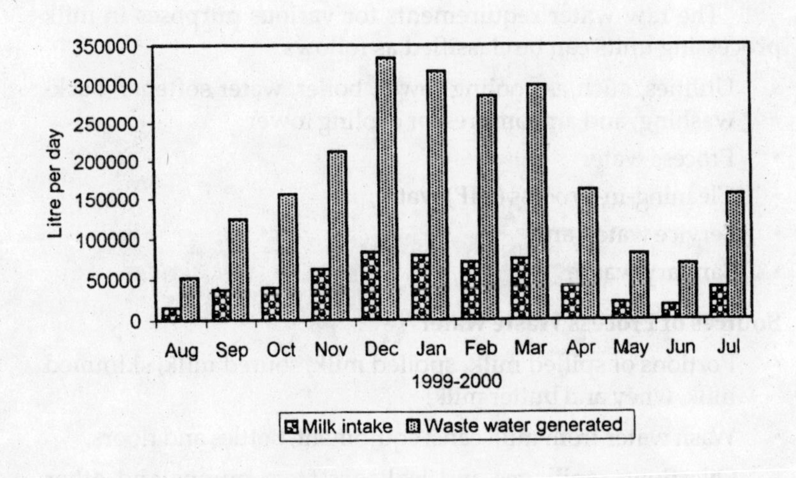

Fig. 3.6 : Monthly milk procurement and waste water generation (Ramnagar Dairy, Varanasi)

Chemicals used in cleaning and sanitation (yearly)

Caustic soda	11500 kg
Soda ash	8000 kg
Bleaching powder	400 kg
Nitric acid	700 Litre
Idophore or Medichlore	120 kg

Ingredients in milk products

Sugar	60 gin L^{-1} of milk product
Common salt	20-25 kg/month
Citric acid	10kg/month

Packaging

Poly pack, butterpaper, butterliner, paneerliner.

Manufacturing Process in Major Products Relevant to Pollution

- Market milk
- Butter

Water Use Pattern

The raw water requirements for various purposes in milk processing units can be classified as follows:

- Utilities, such as cooling tower, boiler, water softener, back-washing, and air compressor cooling tower
- Process water
- Cleaning-in-process (CIP) water;
- Service water; and
- Sanitary water,

Sources of Process Waste water

- Portions of spilled milk, spoiled milk, soured milk, skimmed milk, whey and butter milk;
- Wash water from milk cans, equipment, bottles and floors,
- Overflows, spillages and leakages from pumps and other equipment; and
- Entrainment during evaporation.

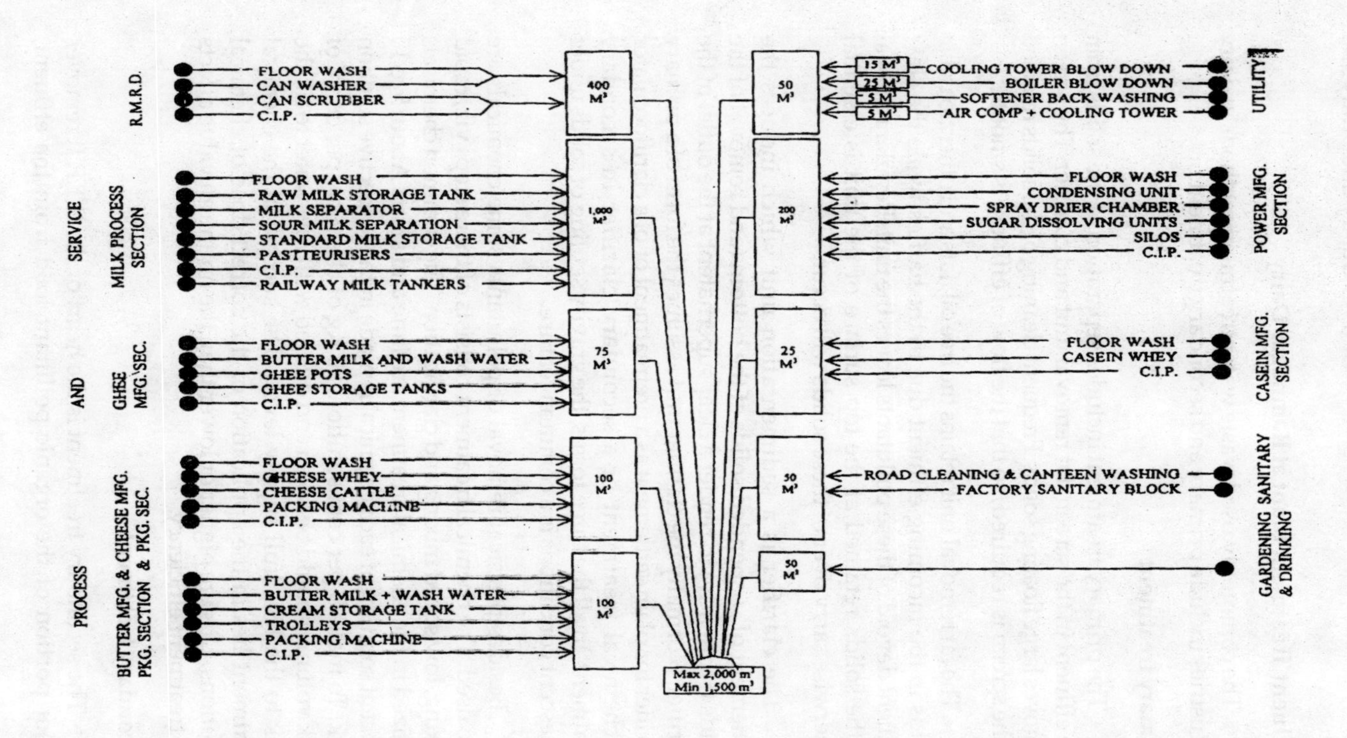
R.M.R.D.
FLOOR WASH
CAN WASHER
CAN SCRUBBER
C.I.P.
400 M³
SERVICE
MILK PROCESS SECTION
FLOOR WASH
RAW MILK STORAGE TANK
MILK SEPARATOR
SOUR MILK SEPARATION
STANDARD MILK STORAGE TANK
PASTTEURISERS
C.I.P.
RAILWAY MILK TANKERS
1,000 M³
AND
GHEE MFG.\SEC.
FLOOR WASH
BUTTER MILK AND WASH WATER
GHEE POTS
GHEE STORAGE TANKS
C.I.P.
75 M³
PROCESS
BUTTER MFG. & CHEESE MFG.
PKG. SECTION & PKG. SEC.
FLOOR WASH
CHEESE WHEY
CHEESE CATTLE
PACKING MACHINE
C.I.P.
100 M³
FLOOR WASH
BUTTER MILK + WASH WATER
CREAM STORAGE TANK
TROLLEYS
PACKING MACHINE
C.I.P.
100 M³
Max 2,000 m³
Min 1,500 m³
UTILITY
COOLING TOWER BLOW DOWN
BOILER BLOW DOWN
SOFTENER BACK WASHING
AIR COMP + COOLING TOWER
15 M³
25 M³
5 M³
5 M³
50 M³
POWER MFG. SECTION
FLOOR WASH
CONDENSING UNIT
SPRAY DRIER CHAMBER
SUGAR DISSOLVING UNITS
SILOS
C.I.P.
200 M³
CASEIN MFG. SECTION
FLOOR WASH
CASEIN WHEY
C.I.P.
25 M³
SANITARY
ROAD CLEANING & CANTEEN WASHING
FACTORY SANITARY BLOCK
50 M³
GARDENING & DRINKING
50 M³

Effluent Treatment Plant at Ramnagar Dairy

The commonly used waste water treatment methods in dairy industries include primary and secondary treatment.

Primary treatment

The primary treatment includes separation of the solids from the effluent in the screen, fat removal unit and clarifier. The screen removes large floating solids. Frequent cleaning of the solids retained at the screen is required, so that the flow of effluent is smooth.

The fat removal unit retains the free oil, fats and other floatable solids in the incoming effluent due to the baffles while the clear effluent devoid of these pollutants leaves the unit. Regular removal of the solids retained at the top surface of the unit is essential otherwise carry over of these solids will occur.

The clarifier is a sedimentation unit which induces the settlement of suspended solids at the bottom (and removal of the solids frequently) to render a clear supernatent at the outlet of the clarifier. Normally the clarifier unit assumes the form of a primary clarifier for plain settlement as a pretreatment or of a clariflocculator in chemical treatment or a secondary clarifier in secondary treatment. In all the three forms, the gravity settling of solids under quiescent conditions is a common feature.

Besides physical removal of pollutants some chemicals are also used. The chemical treatment consists of there steps viz, rapid mixing, low speed mixing and clarification. The common chemicals utilized in the chemical treatment are lime and alum. About 5 to 10 per cent solution of each chemical is made in the respective solution tank. To make 5 per cent solution, 50 Kg of chemical per cu.m. of tank volume should be taken and diluted with raw water to fill the tank to the FSL (full supply level). The success of the chemical treatment lies in the application of the correct dose of chemical solutions. Dosage higher or lower than the optimum value affects the treatment efficiency.

Secondary treatment

The secondary treatment is the heart of the E.T.P. It removes major portion of the organic pollutant load from the effluent, provided the treatment technique is properly understood and it is carefully operated.

Unlike the primary treatment the secondary treatment cannot be put into operation immediately, but it requires a preparation viz, seeding, acclimatization and stabilization before coming to the normal operation stage.

Activated sludge process

The activated sludge process is a modified aeration method. This type of method consists of a holding tank, one or more aeration tanks, and a final sedimentation tank. Aeration is by diffused air from blowers or by mechanical agitation. In the process, the growth of the bacterial floor is achieved in the aeration tank and settles out in the final sedimentation tank. The activated sludge is partly recirculated, either continuously or periodically, to the aeration tank to ensure re-inoculation and to maintain a steady concentration of micro-organisms in the incoming waste water.

Trickling filter

Trickling filter is a bed of graded crushed stone-gravel, or cinders or relatively large size to which waste water is applied on the surface by sprinkling or otherwise. The effluent trickles in a thin film over the surfaces of the filtering medium which have become coated with biological slime. Some of the solids are held by the film. Aerobic bacteria, in the presence of air in the filter, break down the suspended organic solids. Such filters reduce BOD by 78 to 90 per cent.

The most economical solution to obtain 98-99 per cent BOD removal efficiency seems to consist of combining a recirculated trickling filter as a pre treatment and an activated sludge as a second stage.

CHAPTER – IV

MATERIAL AND METHODS

Sampling

Samples were collected twice a day i.e. 8:00 a.m, and 8:00 p.m. at fortnight intervals from the point intake to treatment plant and discharge point of the effluent treatment plant (ET.P.) outside the premises (August 1998 to July 2000). Water samples were collected in triplicate in plastic containers using standard methods of collection. Water samples were brought to the laboratory at 4^{0}C and preservation of samples were done by the addition of 2.5 ml chloroform in 500 ml of water for further analysis of various physico-chemical parameters i.e., Biochemical oxygen demand, Chemical oxygen demand, Total dissolved solids, Total suspended solids, Oil and grease, Nitrogen, Phosphorus, Chloride and Sulphate. The temperature and pH were analysed on the spot.

Analytical Methods

Standard methods for the examination of water and waste water, APHA 1995 was used for analysing various physico-chemical characteristics of dairy waste water.

Physico-Chemical Characteristics

Temperature

The dairy waste water temperature was measured at the influent and effluent discharge point using celsius thermometer expressed in degree centigrade (^{0}C).

pH

An electronic pH meter (CK-701) to the accuracy of 0.05, was used for the measurement of pH. The pH meter was standardized with stock buffers before each reading.

Total Dissolved Solids

In natural waters, the dissolved solids consist mainly of bicarbonates, carbonates, sulphates, chloride, nitrates and possibly phosphates of calcium, magnesium, sodium and potassium with traces of iron, magnesium, sodium and potassium and other substances. Dissolved solids in industrial water are undesirable for many reasons. They form scales, cause foaming in boilers, accelerate corrosion and interfere with the colour and taste of the many finished products.

For its estimation sample is centrifuged or filtered through Whatman No.30 or equivalent filter paper or through a glass fibre filter paper. Evaporate the filtered sample in a tared dish. Porcelain dish is used for sewage and industrial effluents. Dry the residue at 103-10^{0}C or at 179-181^{0}C. The increase in weight of the dish equals the total dissolved solids. It may also be obtained by the difference between the total solids and total suspended solids. Result is expressed as total dissolved solids on drying mg L^{-1}.

Total Suspended Solids

The suspended matter is determined by filtering or centrifuging the sample, drying the residue and determining its weight by difference. Results are expressed as mg L^{-1}.

Filtration method

Gooch crucible containing asbestos mat is used for filtration. Filter a suitable volume of the sample through the crucible by applying suction. The filtration should be carried out at a lower vacuum than that used in the preparation of the mat. Wash the filter with distilled water to remove the soluble salts. Care should be taken to prevent loss of fine asbestos particles during filtration and washing. Dry the crucible in an oven at 103-105°C. The increase in weight equals the total suspended solids.

Biochemical Oxygen Demand

Biochemical oxygen demand (BOD_5) is a test of great value in the analysis of sewage, industrial effluents and polluted water. BOD_5

refers to the quantity of oxygen required by bacteria and other micro-organisms in the biochemical degradation and transformation of organic matter under aerobic conditions. The basic principle underlying the BOD_5 determination is the measurement of the dissolved oxygen content of the sample before and after five days incubation at 20ºC.

To measure the dissolved oxygen content of the water samples, Winklers modified azide method was used. Dissolved oxygen of the sample was measured by precipitating as manganic basic oxide, which was dissolved by concentrated sulphuric acid resulting in the formation of manganic sulphate. This immediately reacts with potassium iodide, already present, liberating iodine which was determined by titrating with sodium thiosuiphate (0.025N) using starch, as an indicator. The chemical reaction involved were as follows :

$$MnSO_4 + 2KOH \rightarrow Mn(OH)_2 + K_2SO_4$$

$$2Mn(OH)_2 + O_2 \rightarrow 2MnO(OH)_2$$
(Manganic basic oxide)

$$MnO(OH)_2 + 2H_2S0_4 \rightarrow Mn(SO_4)_2 + 3H_2O$$
(Manganic sulphate)

$$Mn(SO_4)_2 + 2KI \rightarrow MnSO_4 + K_2SO_4 + I_2$$
(Iodine)

$$2Na_2S_2O_3 + I_2 \rightarrow Na_2S_4O_6 + 2NaI$$

The quantity of iodine liberated during this reaction is equivalent to the quantity of oxygen present in the water sample.

The DO value was calculated by the following formula:

$$\text{Dissolved Oxygen} = \frac{V \times N \times 8 \times 1000}{\text{ml of sample}} \text{ mg L}^{-1}$$

where, V and N are the volume and normality of the titrant respectively.

$$BOD_5 = \frac{DO\left(\text{initial}\right) - DO\left(5 \text{ days incubation}\right)}{\text{Decimal fraction of dilution}} \text{ mg L}^{-1}$$

where,

DO (initial) = Initial DO of the sample water.

DO (5 days incubation) = DO of the sample water after 5 days incubation at 20^0C in dark.

Chemical Oxygen Demand

The dichromate reflux method was used to analyse the COD. The known volume of water sample was refluxed with known volume of potassium dichromate and concentrated sulphuric acid for two hours. The remaining amount of potassium dichromate after completing reflux was titrated with ferrous ammonium sulphate using ferroin indicator. The reaction involved was as follows:

$$K_2Cr_2O_7 + 4H_2SO_4 = K_2SO_4 + Cr_2(SO_4)_3.4H_2O + 3O$$

COD was calculated by following formula–

$$COD = \frac{(A - B) \times N \times 8000}{\text{ml of sample}} \text{ mg L}^{-1}$$

where,

A = ml of ferrous ammonium sulphate used for the blank

B = ml of ferrous ammonium sulphate used for the water sample

N = Normality of ferrous ammonium sulphate

The COD is the quantitative measure of the amount of DO required for the chemical oxidation of carbonaceous materials in waste water using inorganic dichromate salts during a two hour test.

Oil and Grease

The determination of oil and grease includes all the substances that are extractable by the specified solvent. Generally, the substances extractable are, oils, fats and waxes. The results obtained indicates only the non-volatile fraction of these materials.

Solvent extraction method

Oil, grease and other extractable matters are dissolved, in suitable solvent and separated from the aqueous phase. The solvent layer is then evaporated and the residue is weighed as oil and grease.

Calculation

$$\text{Oil and grease}\left(\text{mg L}^{-1}\right) = \frac{\text{Residue in the distilling flask}}{\text{Sample taken for determination}} \times 1000$$

Total Kjeldahl nitrogen

Total Kjeldahl nitrogen is the sum of ammonia nitrogen and organic nitrogen. This does not include nitrite nitrogen and nitrate nitrogen. The classical Kjeldahl method is used to determine the total nitrogen content.

The nitrogen of the organic matter is converted into ammonium sulphate when treated with sulphuric acid in the presence of copper sulphate catalyst. An excess of alkali is then added (to liberate the ammonia from ammonium sulphate) and distilled and the distillate is either treated with Nessler reagent or titrated with standard sulphuric acid after absorbed in boric acid solution.

$$2NH_3 + 2H_3BO_3 \rightarrow 2NH_4^+ \sim + 2H_2BO_3^-$$

$$H_2SO_4 \rightarrow 2H^+ + SO_4^{2-}$$

$$2NH_4^+ + 2H_2BO_3^- + 2H^+ + SO_4^{2-} \rightarrow (NH_4)_2SO_4 + 2H_3BO_2$$

Calculation

Total Kjeldahl nitrogen-N (mg L^{-1})

$$= \frac{1000 \times 0.28 \times \left(\text{ml. } 0.02 \text{ N. } H_2SO^4 \text{ for sample ml. } 0.02 \text{ N. } H_2SO_4 \text{ for blank}\right)}{\text{Sample taken for titration}}$$

Phosphate-P

Stannous chloride method

Ammonium molybdate reacts with phosphate to form molybdophosphoric acid which is reduced to a blue coloured complex 'molybdenum blue' by the addition of stannous chloride.

$$PO_4^{3-} + 12\,(NH_4)_2\,MoO_4 + 24H^+ \rightarrow (NH_4)_3PO_4.12MoO_3 + 21NH_4^+ + 12H_2O$$

$$(NH_4)_3PO_4.12MoO_3 + Sn^{2+} \rightarrow (\text{Molybdenum blue}) + Sn^{4+}$$

A calibration curve is prepared to find out the number of micrograms of P equivalent to the observed optical density of the sample using a spectrophotometer at 690 nm.

Chloride

Chloride is the common anion found in water and sewage. The concentration of chloride in natural waters varies from a few milligrams to several thousand milligrams per litre. Higher concentration of chloride may be due to contamination by brines, sewages or industrial effluents.

Silver Nitrate Method

Silver nitrate reacts with chloride ions to form silver chloride. The completion of reaction is indicated by the red colour produced by the reaction of silver nitrate with potassium chromate solution which is added as an indicator.

$$AgNO_3 + CI \rightarrow AgCI\downarrow + NO_3^-$$

$$2AgNO_3 + KCrO_4\downarrow Ag_2CrO_4\downarrow + 2KNO_3$$

Calculation

$$\text{Chloride as } Cl^-\left(\text{mg } L^{-1}\right) = \frac{(AgNO_3 \text{ solution for sample} - \text{ml. } AgNO_3 \text{ for blank}) \times \text{Normality of } AgNO_3 \times 35.45 \times 1000}{\text{Sample taken for estimation}}$$

If the normality of $AgNO_3$ solution is exactly 0.0282 N, then

$$\text{Chloride as } Cl^-\left(\text{mg } L^{-1}\right) = \frac{(AgNO_3 \text{ solu. for sample} - AgNO_3 \text{ solu. for blak}) \times 1000}{\text{Sample taken for estimation}}$$

Sulphate

Barium chloride colorimetric method was used to analyse the sulphate content of water sample. NaCl-HCl solution and glycerol-ethanol solution were added respectively to the water followed by a pinch of $BaCI_2$. The method is based on the fact that sulphate present in water gets precipitated in acid solution to form white barium sulphate crystals of uniform size. Glycerol-ethanol solution is added as stabilizer. Optical density was taken at 420 nm and sulphate was measured with the help of standard graph of sulphate prepared from known concentration of sulphate solution. The reaction involved was:

$$Na_2SO_4 + BaCl_2 \rightarrow 2NaCI + BaSO_4$$

(White ppt)

Plant Tissue Analysis

Plant tissues were thoroughly washed with tap water. The leaves, stems and roots were separated. Elemental analysis of plant material requires complete breakdown or oxidation of all organic material. The plants were oven dried at 100^{0}C before analysis. This treatment should remove the surface mineral incrustations and thus gives a measure of accurate mineral concentrations. Potasium, calcium and magnesium were determined from the ashed content obtained at 650^{0}C heating. Calcium, sodium, potassium, and magnesium were estimated by using flame photometer. Crude protein content is measured as N x 6.25 (Allen, 1974; Allen et al., 1986).

Total nitrogen in plants

Total nitrogen in aquatic macrophytes was estimated by MicroKjeldahl method given by Peach and Tracey, 1956, Misra, 1968. The air dried (at 80^{0}C) plant sample of 0.5 g weight was digested with concentrated sulphuric acid. Copper sulphate, mercuric oxide, selenium powder were added as catalysts.

Distillation was done in Markham distillation unit and evolved ammonia was collected in boric acid. After adding the Brown crisyl indicator remaining ammonia was titrated with sulphuric acid (N/28). The nitrate content of the plants was calculated by the following formula:

$$\text{Total nitrogen of plant} = \frac{A - B \times N \times 1.4}{S}$$

where,

A = mL of titrant used for sample titration,

B = mL of titrant used for blank,

N = Normality of H_2SO_4

S = weight of the plant sample

Total phosphorus in plants

Chlorostannous reduced molybdophosphoric blue colour method was used to estimate the PO_4-P content of the plants (Jackson, 1962). The air dried plant sample (dried for 12 hrs. at 80^{0}C) was ground and to check volatilization of phosphate, magnesium nitrate was added and was kept in Muffle furnace for 12 hrs. Then it was dissolved in hydrochloric acid, filtered and diluted to a known percentage. pH of the extract (2 ml) was fixed with 2N H_2SO_4 and 4N Na_2CO_3 in the presence of 2,4dinitrophenyl indicator (0.25%). Freshly prepared chlorostannous acid is added which gives blue colour. The colour intensity is equivalent to the phosphate content present in the plant and the intensity is read in terms of optical density at 690 nm on a spectrophotometer.

Statistical Analysis

The data obtained were subjected to different statistical analysis, to determine the significance of variation in the data and to draw valid conclusions. Regression model was also made between COD-BOD.

Analysis of Variance

ANOVA was calculated to test the validity of the data and significance of the variation in the data, between the dairy waste water and among the months.

Correlation Analysis

Correlations between different physico-chemical parameters of dairy waste water (raw and treated) was calculated with the help of computer.

CHAPTER – V

PHYSICO-CHEMICAL PROPERTIES OF DAIRY WASTE WATER

Introduction

The physico-chemical properties of the dairy waste water i.e. raw waste water (before treatment) and treated waste water (after treatment) were analysed at fortnight intervals for a period of two years from August 1998 to July 2000. The monthly variations in different physico-chemical properties of the waste water have been shown in Figs. 5.1 to 5.11.

The statistical analysis was also accomplished. The significance of the variation in the data between the raw waste water and treated waste water during different months for different parameters were calculated using ANOVA test. Correlations between different parameters were also computed.

Quality of Raw and Treated Dairy Waste water

Temperature

Temperature was observed maximum 27.11±0.40°C in May in raw waste water while its minimum value 21.30±0.21°C was observed in the month of December. In case of treated waste water maximum value of it was recorded 25.3±0.50°C in the month of May while minimum value was recorded in December as 19.0±0.08°C (Fig. 5.1). A decreasing trend from summer to winter was noted which may be due to variation in ambient air temperature. Whereas slightly higher temperature of raw waste water in comparison to the treated waste water may be associated with the hot water used during the processing of milk.

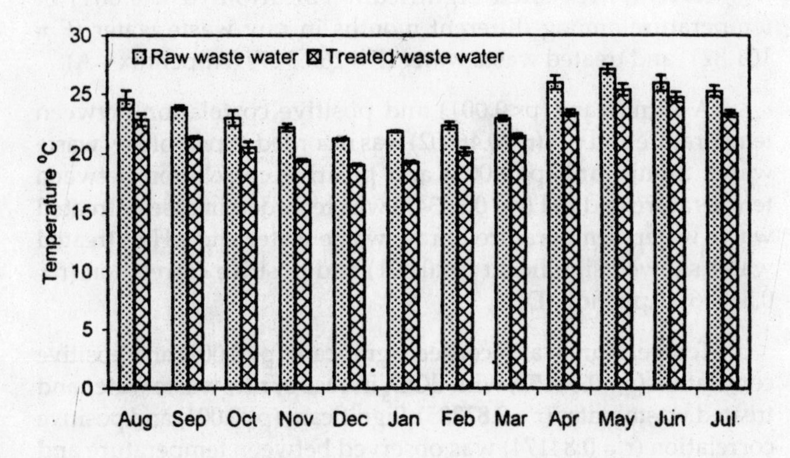

Fig. 5.1 : Monthly variations in Temperature of dairy waste water (Mean values from August 1998 to July 2000; vertical bars indicate 1 ± SE)

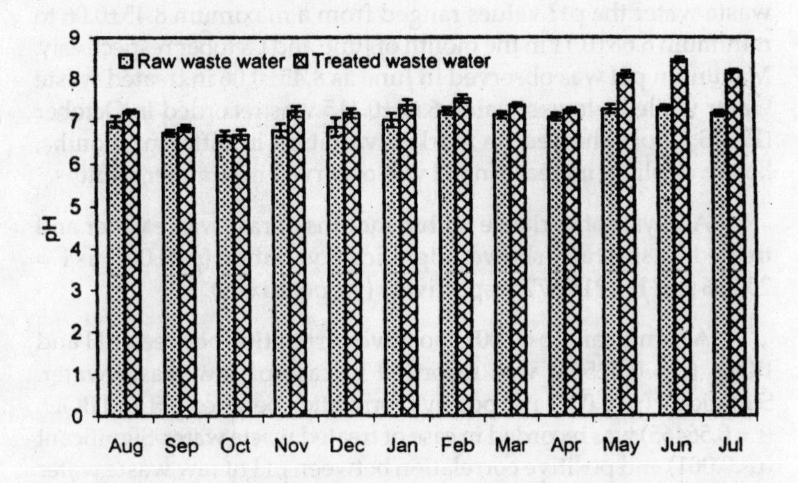

Fig. 5.2 : Monthly variations in pH of dairy waste water (Mean values from August 1998 to July 2000; vertical bars indicate 1 ± SE)

ANOVA revealed significant variation ($p < 0.001$) of temperature among different months in raw waste water ($F = 101.821$) and treated waste water ($F = 267.3 57$) (Appendix – A).

A significant ($p<0.001$) and positive correlation between temperature and pH ($r = 0.46002$) was recorded in case of raw waste water. Significant ($p<0.001$) and positive correlation between temperature and pH ($r = 0,65347$) was recorded in case of treated waste water. Temperature of raw waste water and pH of treated water showed significant ($p<0.001$) and positive correlation ($r = 0.56596$) (Appendix-D).

Temperature was recorded significant ($p<0.001$) and positive correlation ($r = 0.84457$) with BOD_5 in case of raw waste water and treated waste water ($r = 0.87285$). Significant ($p<0.001$) and positive correlation ($r = 0.84174$) was observed between temperature and BOD_5 of raw and treated waste water (Appendix-B).

pH

Maximum pH value was observed in June as 7.31±0.10 and minimum 6.66±0.11 in October in raw waste water. In case of treated waste water the pH values ranged from a maximum 8.45±0.06 to minimum 6.68±0.11 in the month of June and October respectively. Maximum pH was observed in June as 8.45±0.06 in treated waste water while its lowest value 6.68±0.115 was recorded in October (Fig. 5.2). pH showed no marked variation in different months, however, slight increase in pH was observed in summer months.

Analysis of variance within months of raw waste water and treated waste water showed significant variation ($p<0.001$) as $F = 20.336$ and $F = 218.877$ respectively (Appendix-A).

A significant ($p<0.001$) positive correlation between pH and BOD_5 ($r = 0.32585$) was recorded in case of raw waste water. Significant ($p<0.001$) and positive correlation between pH and BOD_5 ($r = 0.56465$) was recorded in case of treated waste water. Significant ($p<0.001$) and positive correlation between pH of raw waste water and BOD_5 of treated waste water ($r = 0.50240$) was observed in study between raw and treated waste water (Appendix-B).

pH was recorded significant ($p<0.001$) and positive correlation ($r = 0.55161$) with COD in case of raw waste water and treated waste water (r = 0.51201). Significant ($p<0.001$) and positive correlation (r = 0.52051) was observed between pH of raw waste water and COD of treated waste water (Appendix-B).

Total dissolved solids

Higher values of TDSs were observed in summer months followed by rainy and winter months. Highest value 1929.04±6.50 mg L^{-1} was observed in May in the case of raw waste water with 1674.46±24.04 mg L^{-1} as lowest value in the month of December (Fig. 5.3).

Maximum TDSs were observed in May as 1643.36±12.12 mg L^{-1} while minimum 1189.28±5.0 mg L^{-1} value was observed in December in case of treated waste water. Marked variation was observed in TDSs values in different months.

ANOVA revealed significant variation ($p<0.001$) in TDSs of raw waste water (F = 199.518) and treated waste water (F = 469.490) within different months.

Significant ($p<0.001$) positive correlation between TDSs and BOD_5 (r = 0.88050) was recorded in case of raw waste water. Significant ($p<0.001$) and positive correlation between TDSs and BOD_5 (r = 0.78312) was recorded in case of treated waste water. Significant ($p<0.001$) and positive correlation in TDSs of raw waste water and BOD_5 of treated waste water (r = 0.77998) was observed in study between raw and treated waste water (Appendix-B).

Significant ($p<0.001$) and positive correlation (r = 0.89768) was recorded between TDSs and sulphate in case of raw waste water and treated waste water (r = 0.87552). Significant ($p<0.001$) and positive correlation (r = 0.86774) was observed between TDSs of raw waste water and sulphate of treated waste water (Appendix-B).

Total suspended solids

Maximum value of TSSs were observed in May while its minimum values were observed in December as 423.23±11.2 mg L^{-1} and 275.72±4.4 mg L^{-1} respectively in raw waste water (Fig. 5.4). In case of treated waste water maximum value of TSSs were observed in May as 95.21±3.33 mg L^{-1} while its minimum value was observed as 60.75±0.60 mg L^{-1} in the month of December.

Decrease in TSSs values were observed from August to December which increases from January to May. TSSs were observed highest in summer months and lowest in winter months.

ANOVA revealed significant variation ($p<0.001$) within months in case of raw waste water ($F = 452.431$) as well as treated waste water ($F = 363.283$) (Appendix-A).

In correlation analysis significant ($p<0.001$) and positive ($r = 0.96092$) correlation was observed between TSSs and temperature of raw waste water (Appendix-B).

TSSs and temperature were observed significant ($p<0.001$) and positively correlated ($r = 0.95541$) in case of treated waste water and in analysis between raw and treated waste water ($r = 0.94370$) (Appendix-B).

TSSs was recorded significant ($p<0.001$) and positively correlated with BOD_5 ($r = 0.88050$) in case of raw waste water and treated ($r = 0.87064$) waste water. Significant ($p<0.001$) and positive correlation was observed ($r = 0.81967$) between TSSs of raw waste water and BOD_5 of treated waste water (Appendix-B).

Oil and grease

Oil and grease was recorded maximum in the month of May as 289.16±4.26 mg L^{-1} and minimum 194.30±3.68 mg L^{-1} in the month of January in raw waste water (Fig. 5.5). Maximum value of oil and grease in case of treated waste water was recorded 5.83±0.10 mg L^{-1} in May and minimum 2.06±0.06 mg L^{-1} in January.

Oil and grease values were observed highest in summer months which decreases in rainy months and lowest in winter months. This trend was recorded in case of raw waste water as well as treated waste water.

ANOVA revealed significant variation ($p<0.001$) in oil and grease values within months in case of raw waster water ($F = 177.774$) and in treated waste water ($F = 822.130$) (Appendix-A).

In correlation analysis significant ($p<0.001$) and positive ($r = 0.84775$) correlation was observed between oil and grease (OG) and nitrogen of raw waste water (Appendix-B).

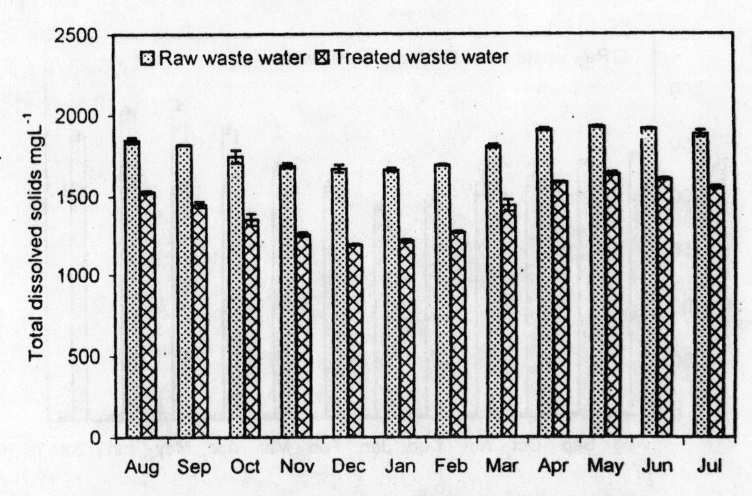

Fig. 5.3 : Monthly variations in Total dissolved solids of dairy waste water (Mean values from August 1998 to July 2000; vertical bars indicate 1 ± SE)

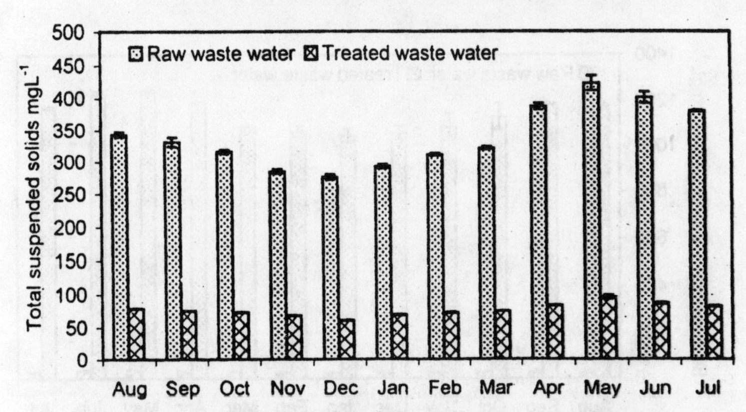

Fig. 5.4 : Monthly variations in Total suspended solids of dairy waste water (Mean values from August 1998 to July 2000; vertical bars indicate 1 ± SE)

OG and nitrogen were observed significant ($p<0.001$) and positively correlated ($r = 0.83051$) in case of treated waste water and in analysis between OG of raw and nitrogen of treated waste water ($r = 0.85533$) (Appendix-B).

OG was recorded significant ($p<0.001$) and positively correlated with BOD_5 ($r = 0.89312$) in case of raw waste water and

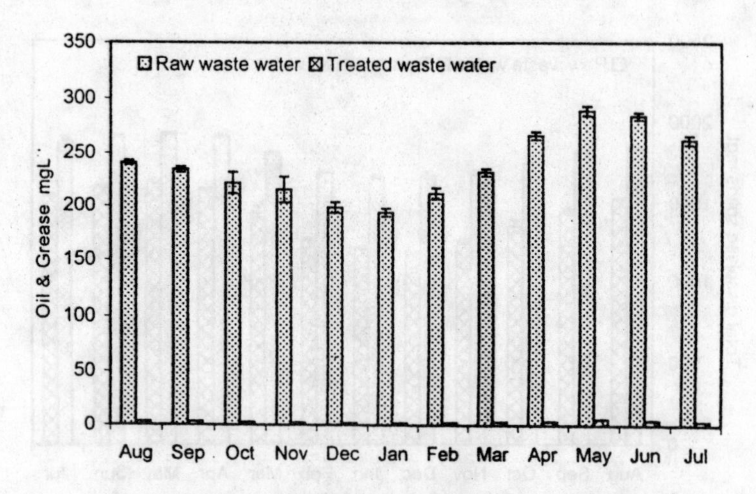

Fig. 5.5 : Monthly variations in Oils and Grease of dairy waste water (Mean values from August 1998 to July 2000; vertical bars indicate 1 ± SE)

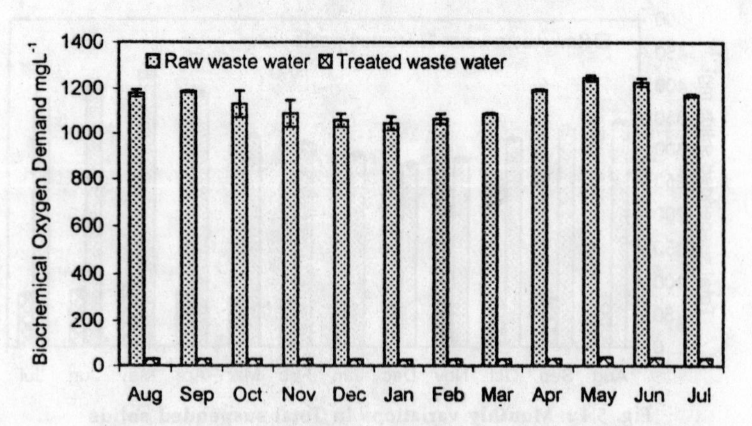

Fig. 5.6 : Monthly variations in Biochemical oxygen demand of dairy waste water (Mean values from August 1998 to July 2000; vertical bars indicate 1 ± SE)

treated (r = 0.953 17) waste water. Significant ($p<0.001$) and positive correlation was observed (r = 0.87883) between OG of raw waste water and BOD_5 of treated waste water (Appendix-B).

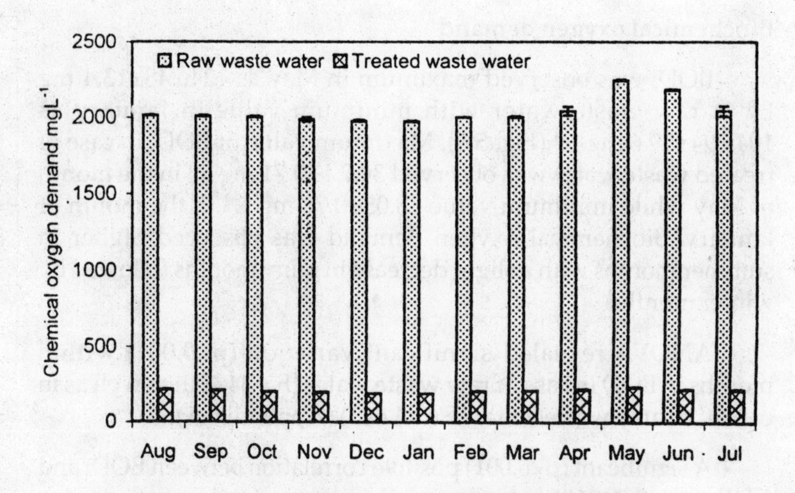

Fig. 5.7 : Monthly variations in Chemical oxygen demand of dairy waste water (Mean values from August 1998 to July 2000; vertical bars indicate 1 ± SE)

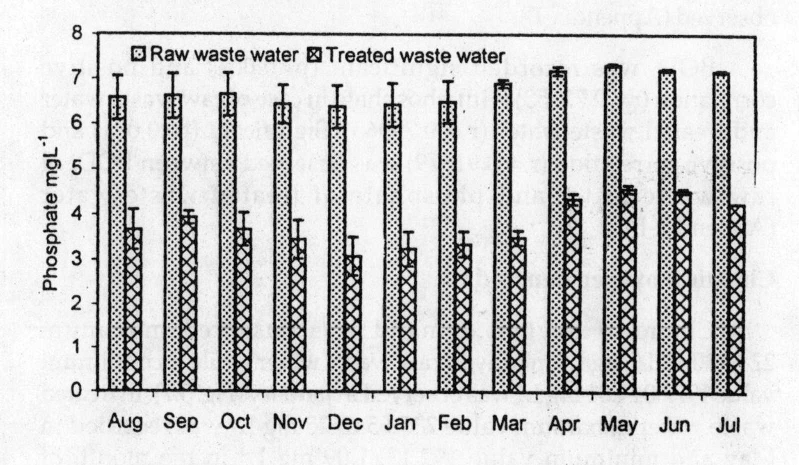

Fig. 5.8 : Monthly variations in Phosphate of dairy waste water (Mean values from August 1998 to July 2000; vertical bars indicate 1 ± SE)

Biochemical oxygen demand

BOD_5 was observed maximum in May as 1246.48±13.4 mg L^{-1} in raw waste water with minimum value in January as 1045.94±29.6 mg L^{-1} (Fig. 5.6). Maximum value of BOD_5 in case of treated waste water was observed 39.2 1±0.71 mg L^{-1} in the month of May while minimum value 28.05±0.76 mg L^{-1} in the month of January. Biochemical oxygen demand was observed higher in summer months with a slight decrease in rainy months followed by winter months.

ANOVA revealed significant variation ($p<0.001$) within months in BOD in case of raw waste water ($F = 34.400$) as well as in case of treated waste water ($F = 94.645$) (Appendix-A).

A significant ($p<0.001$) positive correlation between BOD_5 and COD ($r = 0.74068$) was recorded in case of raw waste water. Significant ($p<0.001$) and positive correlation between BOD_5 and COD ($r = 0.84037$) was recorded in case of treated waste water. Significant ($p<0.001$) and positive correlation between BOD_5 of raw waste water and COD of treated waste water ($r = 0.79683$) was observed (Appendix-B).

BOD_5 was recorded significant ($p<0.001$) and positive correlation ($r = 0.72753$) with phosphate in case of raw waste water and treated waste water ($r = 0.79963$). Significant ($p<0.001$) and positive correlation ($r = 0.91679$) was observed between BOD_5 of raw waste water and phosphate of treated waste water (Appendix-B).

Chemical oxygen demand

Chemical oxygen demand was observed maximum 2245.30±5.16 mg L^{-1} in May in raw waste water while its minimum value 1977.0±4.81 mg L^{-1} was observed in January (Fig. 5.7). In treated waste water maximum value 235.65±3.25 mg L^{-1} was recorded in May and minimum value 192.13±1.09 mg L^{-1} in the month of January.

An increase in values of COD was recorded from January to May. From June, decrease in values of COD was observed with minimum value in January. Seasonally COD was observed highest in summer months and decreases in rainy months and lowest in winter months.

ANOVA revealed significant variation ($p<0.001$) within months in case of raw waste water ($F = 2 19.044$) as well as treated waste water ($F = 132.094$) (Appendix-A).

In correlation analysis significant ($p<0.001$) and positive ($r = 0.84469$) correlation was observed between COD and OG of raw waste water. COD and OG were observed significant ($p<0.001$) and positively correlated ($r = 0.92706$) in case of treated waste water and in analysis between COD of raw and OG of treated waste water ($r = 0.91838$) (Appendix-B).

COD was recorded significant ($p<0.001$) and positively correlated with temperature ($r = 0.79845$) in case of raw waste water and treated ($r = 0.94055$) waste water. Significant ($p<0.001$) and positive correlation was observed ($r = 0.90225$) between COD of raw waste water and temperature of treated waste water (Appendix-B).

Phosphate -P

Phosphorus was observed maximum in the month of May and minimum in December in case of raw waste water as 7.32±0.03 mg L^{-1} and 6.37±0.47 mg respectively (Fig. 5.8). Value of phosphorus in treated waste water was recorded highest 4.6±0.08 mg L^{-1} in May and lowest 3.08±0.42 mg L^{-1} in December.

An increase in the value of P was observed from January to May which starts decreasing from June with minimum value in December. Same trend was observed in case of raw waste water as well as treated waste water.

Analysis of variance among months in case of raw waste water and treated waste water for phosphorus values revealed significant variation ($p<0.001$) $F = 5.759$ and $F = 19.281$ respectively (Appendix-A).

A significant ($p<0.001$) positive correlation between P and TDSs ($r = 0.66985$) was recorded in case of raw waste water. Significant ($p<0.001$) and positive correlation between P and TDSs ($r = 0.84099$) was recorded in case of treated waste water. Significant

($p<0.001$) and positive correlation between P of raw waste water and TDSs of treated waste water ($r = 0.67792$) was observed (Appendix-B).

Total Kjeldahl nitrogen as N

Nitrogen was recorded maximum 77.15±2.65 mg L^{-1} in the month of May in case of raw waste water and minimum 64.0±1.03 mg L^{-1} in January (Fig. 5.9). In case of treated waste water highest value of N was observed 50.09±0.07 mg L^{-1} in May and lowest 40.05±0.04 mg L^{-1} in the month of January.

Seasonally nitrogen was recorded highest in summer season and lowest in winter season. Values of nitrogen were observed increasing from winter to summer season and decreasing from summer to winter season. Same trend was observed in raw as well as treated waste water.

ANOVA revealed significant variation ($p<0.001$) within months in nitrogen in case of raw waste water ($F = 43.308$) and treated waste water ($F = 93.188$) (Appendix-A).

In correlation analysis significant ($p<0.001$) and positive ($r = 0.71617$) correlation was observed between nitrogen and BOD_5 of raw waste water (Appendix-B). Nitrogen and BOD_5 were observed significant ($p<0.001$) and positively correlated ($r = 0.69899$) in case of treated waste water and in analysis between nitrogen of raw and BOD_5 of treated waste water ($r = 0.71617$) (Appendix-B).

Chloride

Maximum value of chloride was observed 144.11±3.70 mg L^{-1} in May and minimum 78.0±2.40 mg L^{-1} in the month of December in case of raw waste water (Fig. 5.10). Chloride was recorded highest in treated waste water in the month of May and lowest in the month of December as 118.66±0.87 mg L^{-1} and 68.45±3.2 mg L^{-1} respectively. Seasonally values of Chloride were observed higher in summer followed by rainy and winter season.

ANOVA revealed significant variation ($p<0.001$) within months in chloride values in case of raw waste water ($F = 893.542$) and treated waste water ($F = 556.675$) (Appendix-A).

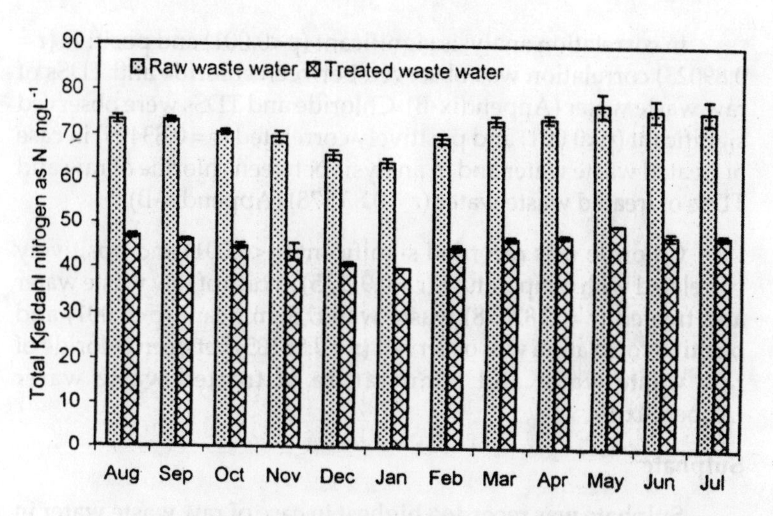

Fig. 5.9 : Monthly variations in Total Kjeldahl nitrogen as N mgL^{-1} of dairy waste water (Mean values from August 1998 to July 2000; vertical bars indicate 1 ± SE)

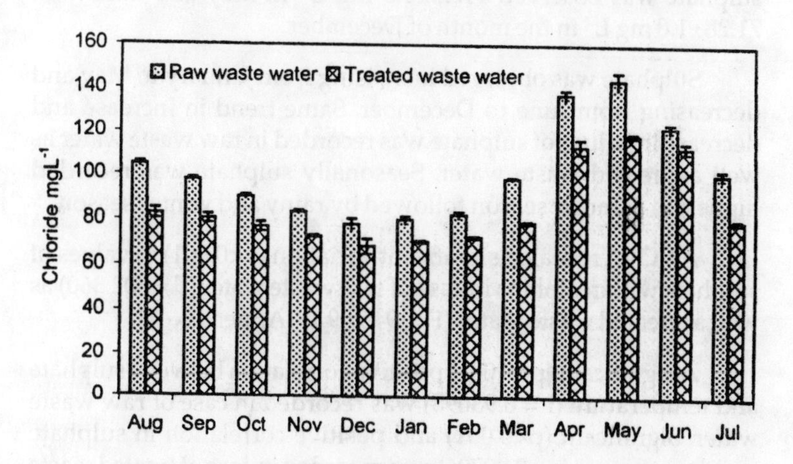

Fig. 5.10 : Monthly variations in Chloride mgL^{-1} of dairy waste water (Mean values from August 1998 to July 2000; vertical bars indicate 1 ± SE)

In correlation analysis significant ($p<0.001$) and positive ($r = 0.89023$) correlation was observed between chloride and TDSs of raw waste water (Appendix-B). Chloride and TDSs were observed significant ($p<0.001$) and positively correlated ($r = 0.83429$) in case of treated waste water and in analysis between chloride of raw and TDSs of treated waste water ($r = 0.83\ 878$) (Appendix-B).

Chloride was recorded significant ($p<0.001$) and positively correlated with temperature ($r = 0.91275$) in case of raw waste water and treated ($r = 0.87078$) waste water. Significant ($p<0.001$) and positive correlation was observed ($r = 0.88105$) between chloride of raw waste water and temperature of treated waste water (Appendix-B).

Sulphate

Sulphate was recorded highest in case of raw waste water in May as 142.25±1.48 mg L^{-1} and lowest in the month of December as 92.54±0.73 mg L^{-1} (Fig. 5.11). In case of treated waste water maximum sulphate was observed 117.0±1.3 mg L^{-1} in May and minimum 71.28±1.0 mg L^{-1} in the month of December.

Sulphate was observed increasing from January to May and decreasing from June to December. Same trend in increase and decrease in values of sulphate was recorded in raw waste water as well as treated waste water. Seasonally sulphate was recorded highest in summer season followed by rainy and winter season.

ANOVA revealed significant variation ($p<0.001$) in values of sulphate within months in case of raw waste water ($F = 302.360$) as well as treated waste water ($F = 945.09\ 1$) (Appendix-A).

A significant ($p<0.001$) positive correlation between sulphate and temperature ($r = 0.93098$) was recorded in case of raw waste water. Significant ($p<0.001$) and positive correlation in sulphate and temperature ($r = 0.89706$) was recorded in case of treated waste water. Significant ($p<0.001$) and positive correlation in sulphate of raw waste water and temperature of treated waste water ($r = 0.91797$) was observed (Appendix-B).

Polynomial regression models for COD-BOD for dairy industry waste water

For many types of waters, it is possible to correlate COD with BOD. This can be very helpful because the COD can be determined in 3 hr, compared to the 5-day duration required for BOD. Once the correlation has been established, COD measurements can be used for the assessment of BOD which will be the good advantage for waste water treatment plant control and operation. BOD-COD relationships for Ganga river waters have been reported by Tiwari et al., 1986. Gajghate and Reddy (1989) have also reported BOD-COD relationship for a few industrial waste waters. A computer programme for polynomial curve fitting has been used to best fit the data given in Figs. 5.6 and 5.7. The results of polynomial curve fitting are shown in Table 5.1.

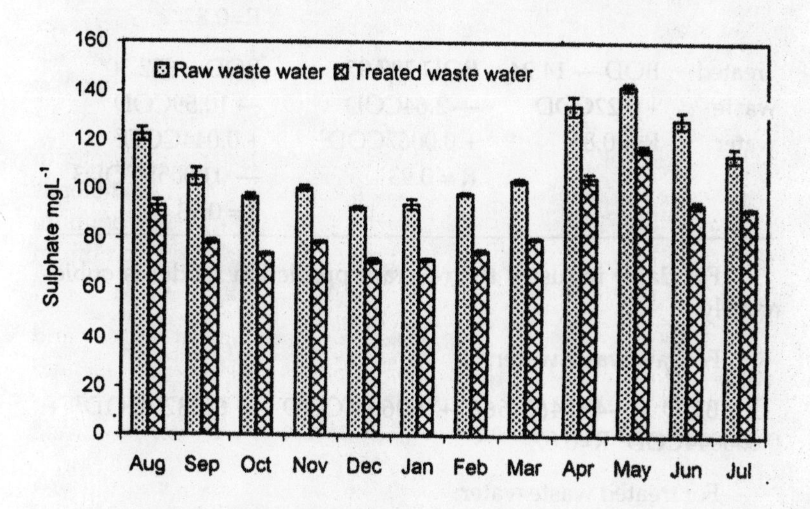

Fig. 5.11 :Monthly variations in Sulphate of dairy waste water (Mean values from August 1998 to July 2000; vertical bars indicate 1 ± SE)

Table 5.1 Polynomial regression models for BOD

Dairy Industry	Linear	Quadratic	Cubic
Raw waste water	BOD=–181.70 + 0.65COD R = 0.74	BOD—24315.74 + 23.62COD —0.0055COD^2 R = 0.83	BOD—431463.666 + 606.04COD —0.2829COD^2 + 0.000044COD^3 R=0.87
Treated waste water	BOD — 14.24 + 0.22COD R = 0.84	BOD 287.97 —2.64COD + 0.0067COD^2 R = 0.93	BOD = 852.30 —10.59COD + 0.044COD^2 —0.00058COD3 R= 0.93

For dairy industry the relevant prediction model is cubic, namely:

For raw waste water:

$$BOD = -431463.666 + 606.04COD - 0.2829COD^2 + 0.000044COD^3 \quad R=0.87$$

For treated waste water:

$$BOD = 852.30 - 10.59COD + 0.044COD^2 - 0.00058COD^3$$

R=0.93

For any given COD value BOD can be calculated from the above equation, with almost negligible percentage error. For dairy waste, it is observed that a cubic model is more appropriate compared to quadratic.

CHAPTER – VI

WASTE WATER TREATMENT BY AQUATIC MACROPHYTES

Introduction

The term 'Aquaculture' means organised culture of aquatic organisms or production of crops of aquatic organisms used as food for humankind. In most kinds of aquaculture the organisms are under man's control until the time of harvest and management practices are aimed at cycling materials and energy at accelerated rates in artificial settings. Aquaculture is also defined as deliberate culture of plants and animals in water, usually for commercial purposes (Avault Jr, 1980).

Nitrogen and phosphorus are considered to be the major inorganic nutrient elements for the growth of plants. However, the excess concentration of either is apt to induce the eutrophication and ultimately leads to pollution of aquatic bodies. The main sources of these nutrients are agricultural drainage, municipal and industrial runoff, fertilizers, domestic waste water and detergents. Several chemical and mechanical techniques have been suggested and tested for upgrading the waste water but they are not satisfactory as in the case of dairy industry nutrients are not removed completely by the secondary treatment method. Recently several aquatic macrophytes growing naturally in polluted aquatic bodies have been used for the removal of nutrients (Sutton and Ornes, 1975, 1977; Cornwell et al., 1977; Wolverton and Mc Donald, 1979; Reddy et al., 1982, 1983, Hauser, 1984; DeBusk et al., 1989; Paul et al., 1989; Tripathi et al., 1990). Many research workers have discussed the nitrogen and phosphorus removal capacity of different aquatic plants (Wolverton et al., 1976; Reddy and DeBusk 1985, Aoyama et al. 1986, DeBusk et al. 1989). Nutrient removal efficiency of a system containing plants will depend on the type of aquatic plant, growth rate of plant, nutrient composition of water, and physico-chemical environment in the water. Aquatic macrophytes which have rapid growth rates and absorb large quantities of nutrients might provide a practical and economic method for tertiary treatment of effluent.

The capacity of vascular aquatic plants to assimilate nutrients from polluted water has been recognized for several years (Rogers and Davis, 1972; Steward, 1970; Boyd, 1976). Vascular aquatic macrophytes such as water hyacinth, duckweed, and cattails, cultured in ponds and reservoirs, offer potential alternatives for treating sewage and industrial effluents (Boyd, 1969; Wooten and Dodd, 1976; Wolverton and Mc Donald, 1979), and agricultural effluents (Reddy et al., 1982). The aquatic macrophytes use solar radiation and thus have a low energy requirement as compared to other methods of tertiary treatment.

Aquatic plants utilize the nutrients and produce large amounts of biomass which can be used for some beneficial purposes. Aquatic macrophyte systems can be effectively used to reduce pollutant levels in water bodies (Boyd, 1969; Lakshman, 1979; Stowell et al., 1981; Reddy et al., 1982) and the biomass used for production of gaseous fuels (Shiralipour and Smith, 1984), feed (Bagnail et al., 1974) fiber (Nolan and Kirmse, 1974) and compost and organic soil amendments (Parra and Hortenstein, 1974).

Studies reported by several researchers (Clock, 1968; Scarsbrook and Davis, 1971; Cornwell et al., 1977) calculate the nutrient removal rates, based on the changes in concentrations at the inflow and outflow of a pond or reservoir. Although these calculations provide information on the nutrient removal efficiency from waste water, they provide very little understanding on the rate of N and P removal in these systems. Many research workers have discussed the nitrogen and phosphorus removal capacity of different aquatic plants (Wolverton et al., 1986; Busk et al., 1989). Culley and Epps suggested that duckweeds might be desirable plant for removing nutrients from sewage effluent. Duckweed would be easier to hardle than water hyacinth because it does not form the dense entangled mat that is characteristic of water hyacinth.

However, most of these studies are restricted to one or very few plants, thus no comparative data among different plants grown under the same environmental conditions are available. The purpose of this study was to evaluate the role of different types of aquatic plants and their combinations in removing N and P from dairy waste water and to establish the role of these plants in improving water quality.

Experimental Plan

Five aquatic macrophytes such as *Eichhornia crassipes, Pistia stratiotes, Lemna minor, Azolla pinnata* and *Spirodela polyrhiza* were selected for the present investigation. Plants were collected from Agrofarm pond of the Banaras Hindu University, Varanasi. Three replicates of each were maintained in 150 Litre glass aquarium. In order to evaluate the actual performance of these selected plants one control set was also taken which was filled only with the waste water with no macrophytes. All the five plants were cultured individually i.e. 100% coverage and in combinations with 50% + 50% coverage of the total surface area of aquarium used for aquaculture. Treated waste water by the conventional treatment plant of Ramnagar dairy was used as a source of waste water.

Roots of all selected plants were washed thoroughly in tap water before the plants were placed in separate glass aquariums. During present investigation 48 aquariums of 150 Litres capacity were used. In each aquarium 95 Litres of waste water was poured.

Experimental sets: with 100% coverage

1. *Eichhornia crassipes*
2. *Pistia stratiotes*
3. *Azolla pinnata*
4. *Lemna minor*
5. *Spirodela polyrhiza*

Exeprimental sets in combinations: (50% + 50% coverage of surface area)

6. *Eichhornia crassipes* & *Lemna minor*
7. *Eichhornia crassipes* & *Azolla pinnata*
8. *Eichhornia crassipes* & *Pistia stratiotes*

9. *Eichhornia crassipes & Spirodela polyrhiza*
10. *Pistia stratiotes & Azolla pinnata*
11. *Pistia stratiotes & Lemna minor*
12. *Pistia stratiotes & Spirodela polyrhiza*
13. *Azolla pinnata & Lemna minor*
14. *Azolla pinnata & Spirodela polyrhiza*
15. *Lemna minor & Spirodela polyrhiza*

Control: Three aquariums were poured with only waste water without any macrophyte and treated as control.

Quality of waste water used for aquaculture

Physico-chemical characteristics of dairy waste water used for aquaculture study are shown in Table 6.1.

Table 6.1: Physico-chemical characteristies of dairy waste water

Temperature (°C)	21.3±0.8
pH	6.97±0.2
TDSs (mg L^{-1})	1420.7±12.2
TSSs (mg L^{-1})	70.4±1.5
Oil & grease (mg L^{-1})	2.7±0.09
BOD_5 (mg L^{-1})	29.1±1.0
COD (mg L^{-1})	209±2.5
Phosphate-P (mgL^{-1})	3.7±0.05
Total Kjeldahl nitrogen-N (mg L^{-1})	43.9±1.2
Chloride (mg L^{-1})	76.4±1.32
Sulphate (mg L^{-1})	79.8±1.1

Calculation

The nutrient removal capacity of each plant species was calculated by using the following formula $C = \frac{I - F}{T}$ (Tripathi et al., 1991).

where C = Nutrient removal capacity of the plant,

I = Initial concentration of nutrient in the water, F = Final concentration of nutrient in the water, and T = time taken for removal.

Duration of the experiments

Study was conducted from 15th Sept. 1999 to 9th Nov. 1999. A constant water level was maintained with the help of distilled water.

Quality of waste water used for aquaculture was analysed at weekly intervals for a period of 56 days because maximum removal was recorded at 56th day of the experiment.

Description of selected aquatic macrophytes

Eichhornia crassipes (Water hyacinth)

Family — Pontederiaceae (Pickerel-weed family)

Eichhornia crassipes is a perennial, naturalised free floating water weed of American origin which occurs widely in ponds, lakes and waterways throughout India. The plant is able to get rooted in the mud when the level of water falls. It grows by means of offsets. Flower formation is also abundant. The stem is horizontal and spongy. The nodes bear rosettes of large aerial leaves and cluster of brown or pinkish adventitious roots. The roots possess root pockets. The leaves are petiolate. The lamina is reniform or rhomboidal in outline. The petioles are inflated. They store air and help the plant to remain afloat on the surface of water. The petioles are strengthened by a sclerenchymatous hypodermis, besides a number of vascular bundles. The roots act as balancers. They contain a central vascular strand. The emerged leaves are thick. They are covered by a water proof cutinised and waxy-coating. The surface is shining. The leaves possess stomata and palisade parenchyma.

Pistia stratiotes (water lettuce)

Family – Araceae (Arum family)

Pistia stratiotes is a monotypic genus of small free floating, perennial aquatic herb of gregarious nature. The plant is found in ponds. It grows by means of offsets. The leaves are sessile. They arise in close spirals and form cupshaped rosettes in the region of nodes. A cluster of adventitious roots arises from each node for providing stability. The roots bear root pockets. The plant is able to remain afloat on the surface of water by means of spongy offsets, minute arenchymatous floats and the horizontally spreading of outer leaves. The leaves are spatulate or obsordate in outline. They possess cuticle and stomata. Mesophyll is differentiated into palisade and spongy parenchyma.

Lemna minor

Family – Lemnaceae (Duckweed family)

Lemna is a small gregarious aquatic plant which is found floating freely on the surface of ponds and lakes. The plant body looks like a thallus. It does not show any distinction of stem and leaves. A plant consists of upto five flat green segment called fronds. Each plant bears a single peltately attached roots. The root is unbranched and hairless. It is devoid of vascular tissues and contains a smooth or winged root pocket (root sheath). The upper surface of the fronds is covered over by an unwettable waxy coating. The stomata are restricted to the surface. Internally the fronds possess spongy parenchyma. Mechanical tissue is lacking.

Azolla pinnata (water fern)

Family - Azollaceae

Azolla is small floating, perennial aquatic with simple unbranched roots. The leaves are small not circinate in venation and dimorphic with 2 lateral, stiff-hairy and floating, and the third submerged, dissected into 8-12 filiform hairy segments, or uniform and alternate. Family available in trade for culture in pools and aquaria.

Spirodela polyrhiza

Family - Lemnaceae (Duckweed family)

Floating or submerged perennial herbs; plants without roots or the roots reduced to unbranched rhizoids; the plant body reduced to a small or minute oval, oblong flat or globose thallus, leafless, often purplish beneath. Species of all genera are cultivated for ornament in poois and aquaria.

In experimental sets containing individual aquatic macrophytes, *Eichhornia crassipes* removed maximum 71.75±0.20% nitrogen. Gradual increase in the removal of nitrogen by *E. crassipes* was observed from day 7th to maximum on day 56th i.e. at the end of the retention period. *E. crassipes* removed maximum nitrogen as compared to control as well as other individual grown plants in aquaculture. Maximum removal of nitrogen by *E. crassipes* may be due to its large size and luxuriant growth in nutrient rich medium.

The comparative study of waste water lagoon with and without water hyacinth was performed by Wolverton and Mc Donald, 1979. They concluded that the pond was almost anaerobic with only traces of dissolved oxygen near the surface in the water hyacinth root zone. This strengthens the removal by *E. crassipes* as deficient level of oxygen in water creates less favorable conditions for nitrification (Reddy, 1981). ANOVA revealed significant variation ($p < 0.001$) in different incubation period ($F = 226406.737$) and non-significant variation among different experimental sets (Appendix-C).

Maximum phosphorus removal 63.21±0.98 per cent was observed in experimental sets containing *E. crassipes*. Regular increase in the removal of phosphorus was observed from first analysis (7th day) to eighth analysis (56th day).

Lesser removal of phosphorus by *E. crassipes* as compared to nitrogen removal may be due to its less requirement, resulting the low up take. Removal of phosphorus in control experimental set (8.44±0.057%) was observed probably due to precipitation with Ca compounds. (Reddy, 1982). ANOVA showed significant variation ($p < 0.001$), in different incubation period ($F = 738.111$) and non-significant variation among different experimental sets (Appendix-C).

Pistia stratiotes removed 64.70±0.40 per cent nitrogen at the end of retention period. An increasing trend in removal of nitrogen was observed from day initial to final day experiment. *Pistia stratiotes* has also been proven as efficient plant for nutrient removal, however, it removed less nitrogen as compared to *E. crassipes*. Statistical variations were found significant among different incubation period ($p<0.001$) (F 65748.624) and non-significant variation among different experimental sets (Appendix-C).

In case of phosphorus removal, *P. stratiotes* removed 60.84±0.86 per cent which was observed at 56th day i.e. at the end of retention period. Removal of phosphorus by *P. stratiotes* was less in comparison to nitrogen removal. An increasing trend in phosphorus removal was observed from initial to final day of experiment. Although significant removal of phosphorus as well as nitrogen was noted by *P. stratiotes*, however, it was slightly lower than that of *E. crassipes*. *P. stratiotes* does not form dense mesh in water body like *E. crassipes* which is an advantage for its use in the aquaculture. ANOVA revealed significant variation ($p <0.001$) among different incubation period ($F = 1379.193$) and non-significant variation among different experimental sets (Appendix-C).

Experimental sets containing *Lemna minor* were observed removing 62.49±0.39 per cent nitrogen which was maximum among small-leaf plants, but less than large-leaf plants (*Eichhornia crassipes* and *Pistia stratiotes*). A gradual increase in the removal of nitrogen was observed from first analysis to eighth analysis (end of retention period). *Lemna minor* was observed having magnificent growth rate, which enhances its capacity to remove nutrients. Harvey and Fox (1973) found that *L. minor* L. would double in frond number every four days when grown in sewage effluent under controlled condition. Its good growth rate supports its use for effluent treatment and biomass production.

Analysis of variance was found significant ($p < 0.001$) among different incubation period ($F = 76789.618$) and non-significant variation among different experimental sets (Appendix-C).

Lemna minor was observed efficient in removing phosphorus (58.80±0.57%) from dairy waste water. Phosphorus removed by *L. minor* was maximum among small-leaf aquatic macrophytes. A gradual increase in the removal of phosphorus was observed from first to eighth analysis. Small leaf plants may be considered as better plants as compared to large leaf plants for aquaculture as these do not form dense mesh and allows greater diffusion of oxygen at air-water interface, dense cover of floating macrophytes (Reddy, 1981) depletes the dissolved oxygen level of the water to < 1 mg L^{-1}, creating less favorable condition for nitrification. ANOVA showed significant ($p < 0.001$) variation among different incubation period (F 6230.336) and among different experimental sets ($p < 0.01$) ($F = 7.539$) (Appendix-C).

Maximum removal of nitrogen by *Azolla pinnata*, 60.13±0.48 per cent was observed at eighth (56th day) analysis. As compared to control experimental sets (13.12±0.02%) it removed a higher percentage of nutrient. Removal in control (with no aquatic macrophytes) experimental sets may be due to nitrification and denitrification processes in water. Similar results in case of control were also observed by Paul et al., 1989.

ANOVA revealed significant variation among different incubation period ($p < 0.001$) ($F = 43412.214$) and non-significant variation among different experimental sets (Appendix-C).

In case of phosphorus removal by *Azolla pinnata* containing experimental sets highest reduction was recorded at 56th day of retention (56.27±0.51%). An increase in the reduction of phosphorus was observed from first analysis to eighth analysis.

Analysis of variance revealed significant ($p < 0.001$) variation among different incubation period ($F = 17241.514$) and non-significant variation among different experimental sets (Appendix-C).

Spirodela polyrhiza also showed significant removal of nitrogen (57.89±0.04 %), which was observed highest on 56th day analysis. Removal of N by *S. polyrhiza* was very high as compared to control (13.12±0.02 %) experimental sets. Gradual increase is the removal of N was observed from first to eighth analysis.

ANOVA showed significant variation among different incubation period ($p < 0.001$) ($F = 63268.984$) and non-significant variation among different experimental sets (Appendix-C).

Spirodela polyrhiza also contributed significantly in reduction of phosphorus from dairy waste water. Maximum phosphorus removal by *S. polyrhiza* 53.32±1.23 per cent was observed at the end of retention period.

An increasing trend in the removal of P was recorded from first to last analysis, which was higher as compared to control (8.44±0.057%) experimental sets (containing no aquatic macrophytes).

Statistical variations were found significant ($p < 0.001$) among different incubation period ($F = 6493.260$) and non-significant variation among different experimental sets (Appendix-C).

In combination study of aquatic macrophytes all the selected plants were cultured in combination 50%+50% coverage of the total surface area of aquarium.

Combination of *Eichhornia crassipes* and *Lemna minor* showed maximum 78.85±0.18 per cent removal of nitrogen and it was observed at the end of retention period i.e. 56th day, in further analysis no increase in removal was observed. Combination of *E. crassipes* and *L. minor* showed great rise in reduction of nutrient as compared to their individual aquaculture experimental sets as well as other combinations. *E. crassipes* and *L. minor* achieved luxuriant growth in nutrient rich medium due to remarkable uptake of nutrients (Tables 6.2, 6.3 and 6.4). *E. crassipes* forms dense mesh which causes anoxic condition in water body, but *L. minor* enhances (at air-water interface) diffusion of oxygen, due to this aerobic condition maintains in the system. Maximum removal in this case may be due to synergistic effects between the two plant species.

Analysis of variance revealed significant ($p < 0.001$) variation among, different incubation period ($F = 395985.122$) and non-significant among different experimental sets (Appendix-C).

Phosphorus removal by *E. crassipes* and *L. minor* combination was observed maximum among all the combinations as well as individual experimental sets studied. Combination of *E. crassipes* and *L. minor* removed 69.37±0.11 per cent of phosphorus. An increasing trend was observed in phosphorus removal by *E. crassipes* and *L. minor* combination from first analysis (7th day) to eighth analysis (5 6th day).

Removal of phosphorus by the combination was recorded less than the removal of nitrogen by it (*E. crassipes* and *L. minor* combination). As compared to control experimental sets sharp rise in reduction of phosphorus by combination of *E. crassipes* and *L. minor* was observed, which was maximum at the end of retention period.

ANOVA revealed significant ($p<0.001$) variation among different incubation period ($F = 196767.030$) and non-significant among different experimental sets (Appendix-C).

Combination of *Eichhornia crassipes* and *Azolla pinnata* showed remarkable reduction of nitrogen from dairy waste water in aquaculture. Removal was observed in increasing order from first analysis to eighth analysis (end of retention period) with maximum value 74.12±0.11 per cent. Efficient removal of nitrogen by the combination of *E. crassipes* and *A. pinnata* may be due to their luxuriant growth in nutrient rich medium and small leaf size of *Azolla pinnata* which allows diffusion of oxygen at air-water interface.

Statistical variations were found significant ($p < 0.001$) among different incubation period ($F = 811092.617$) and non-significant among different experimental sets (Appendix-C).

Combination of *E. crassipes* and *A. pinnata* was observed removing 68.74±1.00 per cent of phosphorus, which was recorded at the end of retention period. Regular increase in the reduction of phosphorus by *E. crassipes* and *A. pinnata* combination was observed.

Reduction of phosphorus was observed less than reduction of nitrogen by this combination, which was in accordance with other combinations as well as individual experimental sets.

ANOVA revealed significant ($p<0.001$) variation among different incubation period ($F = 8618.125$) and non-significant among different experimental sets (Appendix-C).

Combination of *Eichhornia crassipes* and *Pistia stratiotes* was observed removing 51.51±0.22 per cent nitrogen. Removal of nitrogen by this combination was higher as compared to control experimental sets but it was not remarkable in comparison to other combination experimental sets as well as individual experimental sets.

Reduction in nitrogen removal percentage by *E. crassipes* and *P. stratiotes* as compared to other experimental sets (except control experimental sets) was observed may be due to same kind of morphological features. Both the plant species are large-leaf aquatic macrophytes, they covered the aquarium densely and because of having same kind of nutrient requirement too, the larger plant suffered the growth of comparatively short plant i.e. *Pistia stratiotes* or it may concluded as *E. crassipes* and *P. stratiotes* showed antagonism which affected the removal of nutrients from the dairy waste water.

Analysis of variance revealed significant variation among different incubation period ($p < 0.001$) ($F = 122364.699$) and non-significant among different experimental sets (Appendix-C).

Eichhornia crassipes and *Pistia stratiotes* in combination removed 49.89±1.59 per cent of phosphorus which was higher as compared to control, but it was lesser in comparison to all the combinations and individual experimental sets studied.

Removal of phosphorus by this combination was less than the reduction of nitrogen, which was in accordance with other combinations and individual plants containing experimental sets.

An increasing trend in the reduction of phosphorus was recorded from day 7th to 56th day analysis, in further analysis no increase in removal was observed. Maximum removal 49.89±1.59 per cent was observed at the end of retention period. Low values of phosphorus removal by this combination was observed may be due to existence of same niche between the two species as when grown individually both the aquatic macrophytes gave remarkable results.

Analysis of variance revealed significant ($p<0.001$) variation among different incubation period ($F = 3182.559$) and non-significant among different experimental sets (Appendix-C).

Eichhornia crassipes and *Spirodela polyrhiza* in combination removed 67.20±0.23 per cent nitrogen. A gradual increase in the removal of nitrogen was observed in analysis with maximum at the end of retention period.

Removal of nitrogen by individual plants containing experimental sets, *Spirodela polyrhiza* showed least removal but in combination with *E. crassipes* increase in reduction was recorded which was low in value as compared to the individual plants containing experimental sets of *Eichhonia crassipes*.

Statistical variations were found significant ($p < 0.001$) among different incubation, period ($F = 180278.073$) and non-significant among different experimental sets (Appendix-C).

Phosphorus removal by combination of *E. crassipes* and *S. polyrhiza* was recorded maximum 62.83±0.58 per cent at the end of retention period. Phosphorus reduction by *E. crassipes* and *S. polyrhiza* combination was lesser than nitrogen removed by it. Combination removed more phosphorus as compared to the experimental sets containing individually *S. polyrhiza* but less removal was observed as compared to experimental sets containing only *Eichhornia crassipes*.

ANOVA showed significant ($p<0.001$) variation among different incubation period ($F = 14537.806$) and non-significant among different experimental sets (Appendix-C).

Lemna minor in combination with *A. pinnata* cultured experimental sets were observed removing 69.90±0.25 per cent nitrogen. Reduction by this combination among small-leaf macrophytes containing experimental sets was highest. A gradual increase in the removal of nutrient was observed from first to eighth analysis. Both of these plants grow rapidly in nutrient enriched water and hence rapidly remove the nutrients through uptake.

ANOVA showed significant variation ($p < 0.001$) in different incubation period ($F = 208335.202$) and non-significant among different experimental sets (Appendix-C).

Phosphorus removal by combination of *L. minor* and *A. pinnata* was observed maximum at the end of retention period (66.68±0.95%). Phosphorus removal by this combination was comparatively less than the nitrogen removal, similar kind of results were recorded in other combinations too. Comparatively removal of P by combination was higher as compared to control experimental sets.

ANOVA revealed significant ($p < 0.001$) variation in different incubation period ($F = 11906.746$) and non-significant among different experimental sets (Appendix-C).

L. minor and *S. polyrhiza* combination removed 65.28±0.28 per cent nitrogen. Combination gave better results as compared to individually cultured plants of *L. minor* and *S. polyrhiza* in experimental sets. Reduction in N was observed increasing from first analysis to maximum on eighth analysis, in further analysis no increase in removal was observed.

ANOVA revealed significant variation ($p < 0.001$) in different incubation period ($F = 164171.821$) and non-significant among different experimental sets (Appendix-C).

Combination experimental sets with *L. minor* and *S. polyrhiza* were recorded with highest 60.51±0.25% removal of P at the end of retention period. Gradual increase in removal of P was observed in regular analysis.

ANOVA showed significant ($p < 0.001$) variation in different incubation period ($F = 146233.708$) and non-significant among different experimental sets (Appendix-C).

Combination of *Pistia stratiotes* and *Azolla pinnata* in aquaculture removed 67.36±0.28% nitrogen. An increase in removal was recorded during weekly analysis with maximum at the end of retention period. During further analysis no more increase in removal was observed.

ANOVA revealed significant ($p < 0.001$) variation in different incubation period ($F = 240401.261$) and non-significant variation among different experimental sets (Appendix-C).

Combination experimental sets cultured with *P. stratiotes* and *A. pinnata* were observed with 65.89 ±1.2% removal of phosphorus. During weekly analysis, regular increase in removal of P was observed with maximum at the end of retention period. Further analysis does not revealed any increase in nutrient removal. Removal by control experimental sets was (8.44±0.057%) very small as compared to the combination experimental sets.

ANOVA showed significant variation ($p < 0.001$) in different incubation period ($F = 5966.998$) and non-significant among different experimental sets (Appendix-C).

Pistia stratiotes in combination with *S. polyrhiza* cultured experimental sets were recorded removing 65.29±0.28 per cent nitrogen from dairy waste water maximum removal was observed at the end of retention period during weekly analysis. No more increase in removal of N was recorded during further analysis. A large difference in the removal of nutrient by combination and control was observed. Control experimental sets removed 13.12±0.02 per cent nitrogen.

ANOVA revealed significant ($p < 0.001$) variation in different incubation period ($F = 1.2274E + 06$) and non-significant among different experimental sets (Appendix-C).

Combination experimental sets containing *P. stratiotes* and *S. polyrhiza* removed 60.20±0.17 per cent phosphorus. An increasing trend of removal was recorded in weekly analysis with maximum at the end of retention period. Further analysis does not revealed any increase in removal.

ANOVA showed significant ($p < 0.001$) variation in different incubation period ($F = 457197.976$) and significant among ($p < 0.01$) among different experimental sets ($F = 6.992$) (Appendix-C).

Experimental sets cultured with combination of *Pistia stratiotes* and *Lemna minor* showed 68.0±0.85% removal of nitrogen. Higher percentage of N removal was noted in this combination as compared to individually grown plants. An increasing trend in removal was observed from first analysis to eighth analysis. During further analysis no increase in removal was observed.

ANOVA revealed significant ($p < 0.001$) variation in different incubation period ($F = 1.2208E+06$) and non-significant among different experimental sets (Appendix-C).

Combination of *P. stratiotes* and *L. minor* showed 65.54±1.0% removal of phosphorus. Removal by combination of these two plants was more than the removal by these two plants when cultured in experimental sets individually. A regular increase in removal of P was observed with maximum at the end of retention period.

ANOVA showed significant ($p < 0.001$) variation in different incubation period ($F = 10921.49$) and non-significant among different experimental sets (Appendix-C).

In combination study of *A. pinnata* and *S. polyrhiza* 64.84±0.07% nitrogen removal was recorded at the end of retention period, which was maximum. During weekly analysis an increasing trend in removal was observed from first to eighth analysis. Further analysis does not revealed increase in removal of nitrogen.

ANOVA revealed significant ($p < 0.001$) variation in different incubation period ($F = 2323.764$) and non-significant among different experimental sets (Appendix-C).

Phosphorus removal by combination study of *A. pinnata* and *S. polyrhiza* was observed highest 58.57±1.1% at the end of retention period (56th day), during further analysis no increase in removal was observed.

ANOVA showed significant ($p < 0.001$) variation in different incubation period ($F = 8955.903$) and non-significant among different experimental sets (Appendix-C).

Nutrient content in plant tissues cultured in dairy waste water

Nitrogen and phosphorus contents in plant tissues grown in dairy waste were also analysed to observe the changes. In the aquaculture experimental sets for individual plant species, maximum increase in total nitrogen was recorded in the case of *Eichhornia crassipes* as 58.23% followed by *Pistia stratiotes* 29.20%, *Lemna minor* 28.37%, *Azolla pinnata* 17.76% and minimum in the case of Spirodela polyrhiza with 13.97% (Table 6.2).

Among combination experimental sets, *Eichhornia crassipes* and *Lemna minor* combination showed 59.23% increase in nitrogen in *E. crassipes* and 31.38% increase in *Lemna minor*. Total nitrogen content in *E. crassipes* increased by 58.43% and 18.07% in *Azolla pinnata* in combination study (Table 6.2). Increase in total nitrogen content in plant tissue was recorded 52.6 1% in *E. crassipes* and 25.00% in *P. stratiotes* in combination experimental sets.

Total nitrogen content increases in plant tissues by 56.60% in *E. crassipes* and 14.07% in *S. polyrhiza* in combination experimental sets (Table 6.2). Among combination experimental sets cultured with *Lemna minor* and *Azolla pinnata* an increase of 29.80% and 17.49% respectively was recorded in total nitrogen content in plant tissues. Combination of *L. minor* and *S. polyrhiza* also showed increase in nitrogen content at the end of retention period as 29.40% and 13.04% in *L. minor* and *S. polyrhiza*, respectively (Table 6.2).

An increase in total nitrogen of 23 .90% in *P. stratiotes* and 16.12 % in *A. pinnata* was recorded at the end of retention period in combination experimental sets. An increase of 24.56% in total nitrogen content was observed in *P. stratiotes* and 11.34% in *S. polyrhiza* in combination study. In the case of *P. stratiotes* and *L. minor* an increase of 24.94% and 29.50% was noted respectively. Similarly an increase of 13.71% and 15.73% was noted in the case of *S. polyrhiza* and *A. pinnata* respectively (Table 6.2).

Table 6.2 : Changes in nutrient content in plant tissues grown in dairy waste water (Total nitrogen-N gKg-1) (±1SE)

		Initial	*Final (56th day)*	*% increase*
Individual study	*Eichhornia crassipes*	15.11 ± 0.01	23.91 ± 0.03	58.23
	Pistia stratiotes	18.32 ± 0.01	23.67 ± 0.25	29.20
	Lemna minor	20.30±0.20	26.06±0.15	28.37
	Azolla pinnata	22.69±0.43	26.72± 0.14	17.76
	Spirodela polyrhiza	19.40 ± 0.40	22.11 ± 0.28	13.97
Combination study	*E. crassipes + L. minor*			
	E. crassipes	15.11 ± 0.01	24.06± 0.07	59.23
	L. minor	20.30± 0.20	26.67±0.25	31.38
	E. crassipes + A. pinnata			
	E. crasssipes	15.11 ± 0.01	23.94± 0.05	58.43
	A.pinnata	22.69±0.43	26.79±0.08	18.07
	E. crassipes + P. stratiotes			
	E.crassipes	15.11±0.01	23.0 ±0.06	52.61
	P. stratiates	18.32±0.01	22. 0±0.10	25.00
	E. crassipes + S. polyrhiza			
	E. crassipes	15.11 ± 0.01	23.67± 0.23	56.60
	S. polyrhiza	19.40 ± 0.40	22.13 ± 0.23	14.07
	L. minor + A. pinnata			
	L. minor	20.30 ± 0.20	26.37 ± 0.06	29.80
	A. pinnata	22.70 ± 0.42	26.67 ± 0.06	17.49
	L. minor + S. polyrhiza			
	L. minor	20.30± 0.20	26.27±0.15	29.40
	S. polyrhiza	19.40 ± 0.40	21.93 ± 0.21	13.04
	P. stratiotes + A. pinnata			
	P. stratiotes	18.32± 0.01	22.70± 0.26	23.90
	A. pinnata	22.70 ± 0.43	26.37 ± 0.15	16.12
	P. stratiotes + S. polyrhiza			
	P. stratiotes	18.32± 0.01	22.82±0.13	24.56
	S. polyrhiza	19.40 ± 0.40	21.60 ± 0.36	11.34
	P. stratiotes + L. minor			
	P. stratiotes	18.32± 0.01	22.89± 0.21	24.94
	L. minor	20.30 ± 0.20	26.30 ± 0.10	29.50
	S. polyrhiza + A. pinnata			
	S. polyrhiza	19.40±0.40	22.06±0.11	13.71
	A. pinnata	22.70 ± 0.43	26.27 ± 0.06	15.73

Table 6.3: Changes in nutrient content in plant tissues grown in dairy waste water (Total phosphorus-P gKg-1) (±1SE)

		Initial	*Final (56th day)*	*% increase*
Individual study	*Eichhornia crassipes*	3.43 ± 0.03	5.83 ± 0.04	69.97
	Pistia stratiotes	5.68 ± 0.04	7.58 ± 0.07	33.45
	Lemna minor	6.86 ± 0.05	9.29 ± 0.09	35.42
	Azolla pinnata	6.89 ± 0.09	9.63 ± 0.04	39.77
	Spirodela polyrhiza	6.62 ± 0.21	8.13 ± 0.06	22.80
Combination study	*E. crassipes + L. minor*			
	E. crassipes	3.43 ± 0.03	5.87 ± 0.04	71.14
	L. minor	6.86 ± 0.05	9.38 ± 0.01	36.73
	E. crassipes + A. pinnata			
	F. crasssipes	3.43 ± 0.03	5.83 ± 0.01	69.97
	A. pinnata	6.89 ± 0.09	9.65 ± 0.03	40.05
	E. crassipes + P. stratiotes			
	E. crassipes	3.43 ± 0.03	5.75 ± 0.04	67.63
	P. stratiotes	5.68 ± 0.04	7.48 ± 0.03	3.1.70
	E. crassipes + S. polyrhiza			
	E. crassipes	3.43 ± 0.03	5.81 ± 0.01	69.00
	S. polyrhiza	6.62±0.21	8.20±0.00	23.87
	L. minor + A. pinnata			
	L. minor	6.86 ± 0.05	9.32 ± 0.03	35.86
	A. pinnata	6.89 ± 0.09	9.63 ± 0.02	39.77
	L. minor + S. polyrhiza			
	L. minor	6.86 ± 0.05	9.29 ± 0.01	35.42
	S. polyrhiza	6.62 ± 0.21	8.03 ± 0.06	21.29
	P. stratiotes + A. pinnata			
	P. stratiotes	5.68 ± 0.04	7.56 ± 0.05	33.09
	A. pinnata	6.89 ± 0.09	9.61 ± 0.01	39.47
	P. stratiotes + S. polyrhiza			
	P. stratiotes	5.68 ± 0.04	7.59 ± 0.01	33.62
	S. polyrhiza	6.62 ± 0.21	8.03 ± 0.06	21.30
	P. stratiotes + L. minor			
	P. stratiotes	5.68 ± 0.04	7.61 ± 0.01	33.80
	L. minor	6.86 ± 0.05	9.37 ± 0.06	36.59
	S. polyrhiza + A. pinnata			
	S. polyrhiza	6.62 ± 0.21	7.97 ± 0.06	20.40
	A. pinnata	6.89 ± 0.09	9.53 ± 0.06	38.31

Table 6.4 : Chemical composition of Aquatic macrophytes grown in dairy waste water (%)

Parameters	*Eichhornia crassipes*		*Pistia stratiotes*		*Lemna minor*		*Azolla pinnata*		*Spirodela polyrhiza*	
	Initial	*After 56 days*	*Initial*	*After 56 days*	*Initial*	*After 56 days*	*Initial*	*After 56 days*	*Initial*	*After 56 days*
Calcium	1.83	2.89	1.42	2.10	1.37	1.80	1.36	1.82	1.30	1.67
Potassium	3.75	4.35	3.0	4.11	2.97	3.95	2.94	3.91	2.41	3.25
Sodium	0.61	1.27	0.48	0.90	0.23	0.65	0.26	0.67	0.20	0.60
Magnesium	0.90	1.23	0.71	0.97	0.49	1.01	0.50	1.05	0.40	0.89
Ash	18.59	19.73	14.59	15.13	12.10	12.63	13.93	14.90	12.0	12.67

Table 6.5 : Composition of selected crop plants used as animal feed in India and aquatic macrophytes grown for dairy waste water treatment

Crop plants	*Crude protein (%)*	*Ca (%)*	*P (%)*	*Mg (%)*	*Na (%)*	*K (%)*
Barseem*	14.34	7.75	0.18	0.12		
Padd straw*	3.42	17.87	0.07	-	-	-
Dry maize stock*	11.10	1.94	0.41	-	-	-
Wheat bran*	11.39	9.90	0.82	-	-	-
F. crassies **	15.03	2.89	0.58	1.23	1.27	4.35
*L. minor***	16.60	1.80	0.93	1.01	0.65	3.95

* K.C. Sen and S.N. Ray (1971) Nutritive Value of Indian Cattle Feed, 6th ed., Indian Council of Agricultural Research

** Data produced during present investigation.

Nutrient content analysis in plant tissues for phosphorus, for 100% coverage, *Eichhornia crassipes* showed 69.97% increase while grown alone followed by 33.45% increase in *P. stratiotes*, in *Lemna minor* 35.42%, in *Azolla pinnata* 39.77% and 22.80% in *Spirodela polyrhiza* (Table 6.3).

Phosphorus content analysis in plant tissues under combination experimental sets *E. crassipes* and *L. minor* combination showed 71.14 per cent increase in *E. crassipes* and 36.73% increase in *L. minor*. Whereas *E. crassipes* and *A. pinnata* combination showed 69.97% increase in *E. crassipes* and 40.05% in *A. pinnata* (Table 6.3).

E. crassipes with 67.63% and *P. stratiotes* with 37.70% increase in phosphorus were recorded in combination. Similarly, *E. crassipes* and *S. polyrhiza* combination showed 69.0% and 23.87% increase in *S. polyrhiza*. In *L. minor* and *A. pinnata* combination 35.86% increase in *L. minor* and 39.77% in *A. pinnata* were also recorded.

In case of *L. minor* and *S. polyrhiza* combination about 35.42% and 2 1.29% increase were noted respectively. *Pistia stratiotes* and *Azolla pinnata* were recorded with 33.09% increase in P in *P. stratiotes* and 39.47% in *A. pinnata* (Table 6.3).

In combination experimental sets containing *P. stratiotes* and *S. polyrhiza* an increase in P content were observed as 33.62% in *P. stratiotes* and 21.3% in *S. polyrhiza*.

Similarly, 33.80% in *P. stratiotes* and 36.59% in *L. minor* were noted in their combination study. *Spirodela polyrhiza* and *Azolla pinnata* combination showed 20.40% increase in *S. polyrhiza* and 38.3 1% in *A. pinnata* (Table 6.3).

Analysis of chemical composition of aquatic macrophytes grown in dairy waste water during present investigation revealed that plant tissues (leaf and stem) of *E. crassipes*, *P. stratiotes*, *L. minor*, *A. pinnata*, and *S. polyrhiza* indicated large amount of mineral nutrients, before aquaculture treatment and after treatment (Table 6.4). The results revealed that mineral nutrients were increased in plant tissues and found maximum at 56 days of retention period. Calcium content increased 1.83% to 2.89% in *E. crassipes* and in *L. minor* from 1.37% to 1.80% after treatment. In *P. stratiotes*, calcium increased from 1.42% to 2.10%, in *A. pinnata* from 1.36% to 1.82% and in *S. polyrhiza* from 1.30% to 1.67%.

Increase in potassium was recorded from 3.75% to 4.35% in *E. Crassipes. L. minor* was observed with increase in potassium from 2.97% to 3.95% after treatment. In analysis for *P. stratiotes*, potassium increased from 3.0% to 4.11%, in *A. pinnata* it increased from 2.94% to 3.91% and finally in *S. ployrhiza* from 2.41% to 3.25% at the end of retention period (Table 6.4).

CHAPTER – VII
GENERAL DISCUSSION

Introduction

The environmental pollution as defined by the Environment Protection Act (1986) is "the presence of any solid, liquid or gaseous substances present in the environment in such a concentration, as may be or tend to be injurious to environment." Water is one of the constituent of environment, which is a vital resource for all forms on the earth, but unplanned industrialization, lack of ecological education and modernization have resulted into degradation of aquatic ecosystems. The pollutants entering into an aquatic ecosystem are of two types — non-degradable pollutants, which do not degrade or degrade very slowly in the aquatic environment such as aluminium, mercury salts, DDT, long chain phenolic chemicals. The removal of such pollutants is virtually difficult. Second type of pollutants are biodegradable pollutants. This includes domestic and industrial wastes that can be decomposed by natural processes. These create problem only when the input exceeds into the aquatic ecosystems beyond their decomposition capacity. Thus, pollution by toxic non-degradable substances can not be easily controlled, whereas, the pollution by degradable substance is technically solvable by a combination of mechanical, chemical and biological treatments.

The development of a new type of resource saving and energy saving technical system in the field of water treatment is awaited with the increasing concern on the world wide eutrophication problem caused by degradable pollutants, in lentic and lotic water sources. Recently, the phenomenon of pollutants accumulation by aquatic macrophytes has become a subject of interest for the scientists engaged in biological treatment of water.

Attempts have been made to put the voracious appetite of the naturally occurring aquatic macrophytes to some beneficial uses, by using them in the treatment of waste water. Such plants on the one hand are useful to remove the nutrients from the aquatic bodies and on the other hand produce large amounts of biomass which can be used for some beneficial purposes, such as the production of biogas (Shiralipour and Smith, 1984), fibre (Nolan et al, 1974) compost and organic soil amendments (Parra and Hortenstein, 1974).

The present research plan was designed to analyse the physico-chemical properties and to evolve effective and economic biological treatment method for the treatment of the dairy waste water. Results of the present investigation are compared with the water quality standards, given by various agencies as World Health Organization (WHO), Indian Standard Institution (ISI), Indian Council of Medical Research (ICMR) and US Public Health Service (USPHS). In order to justify the significance of the results and to assess the correlation between various properties, analysis of variance and correlation analysis have also been computed.

Milk procurement and waste water generation

Ramnagar dairy (F.B.D. Ramnagar, Varanasi) has a capacity for milk processing of about 1 lakh Litre/day. Milk collection was recorded higher in winter months depending on the availability of it in the village societies (milk is collected from village societies and then it is send to factory). Lowest procurement in August was recorded due to festivals as milk is collected less in the societies because of consumption in near markets. Waste water generation was recorded highest in December and lowest in August.

Usually three to four times of waste water is generated in handling of milk. The liquid waste originates from manufacturing process, utilities and service sections. Generally, the total volume of waste water generation per unit of milk processed varies from unit to unit. However, an attempt has been made to have an average value of water consumption per unit of milk processed. The raw water Dairy factory uses cold and hot water for milk processing and normal water requirement for various purposes in milk processing units can be classified as utilities, process water, cleaning in process (CIP) water, service water, and sanitary water.

Presently, about 12 MLD of milk is processed daily in the country and an average production of 42 MLD waste water (N.D.D.B., 1993). Efficient waste reduction system will not only reduce the effluent volume, but will also reduce the intensity of treatment required. In Ramnagar dairy conventional treatment plant is established to treat the raw waste water generated from various processing sections.

The quality of raw and treated waste waters of Ramnagar dairy were analysed for a period of two years i.e. August 1998 to July 2000 at fortnight intervals and the results are discussed as follows:

Quality of raw and treated waste water

1. Temperature

Study of waste water temperature is a significant aspect of characterization in case of industrial effluents. Temperature is one of the important factor that affects the biochemical and chemical reactions occurring in the water thus altering its nature. Solubility and diffusion of gases also depends upon the variation in temperature. According to Warren (1971), temperature of aquatic systems usually does not exceeds 37°C. When industrial effluents are produced at high temperature, their temperature is cooled down before discharging into any aquatic bodies to avoid stress due to high heat in natural streams and rivers. The temperature fluctuation in the present study depended upon the ambient temperature and also on the surface area exposed directly to the atmosphere. Seasonally, water temperature was highest in summer followed by rainy and winter season. The results were similar to the reports of UPPCB and CPCB reports (1998).

Dairy factory uses cold and hot water for milk processing and normal water for cleaning and other purposes. Temperature of treated waste water was observed slightly lower as compared to the raw waste water, which may be due to aerators used in effluent treatment plant. When treated waste water comes in contact with ambient air, temperature changes accordingly. Temperature of raw and treated waste water were found within the permissible limits.

Table 7.1 : Water quality standards for inland waters

Parameter	*USPHS*	*ISI*	*WHO*	*ICMR*
Temperature °C	±	40.0	-	-
Transparency cm	-	-	-	-
EC Sm^{-1}	0.03	0.075	-	-
pH	6.0-8.5	6.5-8.5	7.0-8.5	6.5-9.2
DO mg U	>4.0	>5.0	-	-
BOD mg L^{-1}	+	<3.0	-	-
COD mg L^{-1}	-	<20.0	-	-
Chloride mg L^{-1}	250	250	200	250
Alkalinity mg L^{-1} $CaCo_3$	-	-	-	81-120
Free CO_2 mg U'	-	-	-	
Nitrate mg U'	10.0	50.0	45.0	20.0
Phosphate mg U'	0.1	-	-	-
Sulphate mgU	250	150	200	200
Total hardness mg U' $CaCo_3$	500	300	100	300
Total solids mg U	500	-	500	-
Calcium mg U'	100	75	75	75
Magnesium mg U'	-	30	-	50
Potassium mg L^{-1}	-	-	-	20
Sodium mg U'	-	-	50	-

- Not available.

± No consensus on a single numerical value which is universally acceptable

pH

pH is measure of the relative acidity or alkalinity of water. It is negative logarithm of H^+ ions. All chemical and biochemical reactions are influenced by the pH, as specific reactions takes place at the specific pH values. According to Warren (1971), the pH value of natural aquatic ecosystems varies in the range of 6.5 to 8.5, but in present study dairy waste water pH varied from 6.68-8.45 because of the cleaning processes and due to amount of handling of milk. Inspite of the seasonal change, generally pH values varied in the narrow range. Seasonally, higher values were noted during summer followed by rainy and winter season months.

Maximum pH was observed in June. The major factor affecting the pH of dairy plant waste water may be the cleaning compounds. Generally, pH and temperature of dairy wastes were noted higher than those of domestic wastes.

Total dissolved solids

'Solids' is a term applied to all matters except the water which contained liquid materials, and thus the definition of solids refers to the matter that remains as residue upon evaporation after drying at a definite temperature. The amount and nature of dissolved and undissolved matters occurring in liquid material vary considerably. Determination of dissolved and undissolved matter is made with filtered and unfiltered portions of sample.

In natural waters, the dissolved solids consist mainly of bicarbonates, carbonates, sulphates, chlorides, nitrates and possibly phosphates of calcium, magnesium, sodium and potassium with traces of iron, manganese and other substances.

The proportion of non-organic solids in milk is about 13%, which suggests that most of the suspended solids in dairy industry waste waters are of a dairy food origin. An increase of non-organic matter in the total solids reflects a contribution of non-organic material from detergents, sanitizers and lubricants. Dairy wastes generally have low concentrations of settleable solids (material removed by gravity settling) although wastes from casein operations, with high concentrations of fines and wastes from drying plants may be exceptions. Washings from the exterior of tanker trucks and from yards may also contribute to settleable solids concentrations.

Dissolved solids in industrial waste waters are undesirable for many reasons. They form scales, cause foaming in boilers, accelerate corrosion and interfere with the colour and taste of the many finished products. It's relevance to land application systems lies mainly in possible salinity damage to crops and or soil structure and degradation of groundwater. In general, TDSs level lower than 500 mg L^{-1} pose no problem. TDSs were observed maximum in summer season followed by rainy and winter. During present investigation TDSs in treated waste water ranged from 1189.28±5.0 mg L^{-1} 1643.36±12.12 mg L^{-1}. As compared with the standards given by different agencies, values were found beyond the prescribed limits.

Total suspended solids

Disposal of sewage and industrial effluents contributes suspended matter to rivers and streams. Suspended solids are very objectionable in rivers for many reasons. Suspended solids containing much organic matter may cause putrefaction and consequently the streams may be devoid of dissolved oxygen. Mineral and organic suspended matter can lead to siltation and if stream bed is blanketed, destruction of plant and animal life may occur. Further, abrasive materials present in the suspended matter cause not only erosion of metallic surface but also cause injury to fish and can render the stream bed unsuitable for spawning.

Cairns (1968), observed the ecological effects of suspended solids as blanketing action of sedimentation and loss of light penetration etc. In case of land disposal clogging of soil pores can become a problem where there is overloading. It is, therefore, necessary to have a restriction on suspended solids load. Solids loading at the higher end of range may require rest and reaeration periods as well as periodic tillage to restore soil permeability.

TSSs were observed maximum in summer followed by rainy and winter months. The reported suspended solids concentrations in raw dairy wastes vary widely. The suspended solids in dairy wastes are mainly organic in nature, volatile suspended solids, as a percentage of the total suspended solids range from 68% to 98%

with an average of 85% (Kearney, 1973). Casein losses in the whey, wash-water and floor effluents are readily determined by measuring the concentration of suspended solids (casein fines).

During present study the TSSs values ranged from 275.72±4.40 mg L^{-1} to 423.23±11.20 mg L^{-1} in raw waste water which was reduced 60.75±0.60 mg L^{-1} to 95.2 1±3.33 mg L^{-1} after the treatment by conventional treatment plant. Values when compared with the various standards were found higher than the prescribed limit.

Oil and Grease

The determination of oil and grease includes all the substances that are extractable by the specified solvent. Generally, the substances extractable are oils, fats and waxes. The results obtained indicates only the non-volatile fraction of these materials.

The floating oil and grease values make no distinction between petroleum based oils and edible oils and fats. The former are much more difficult to treat biochemically than edible oils and fats. It should be noted that milk fat is highly biodegradable as illustrated by its high BOD^5 value, high BOD_5/COD ratio and high BOD_5 rate constant [0.31 (Brown and Pico, 1980)].

Lipids degrade very slowly because it is difficult for bacteria, to attack hydrophobic globules. Consequently, rest period for fat, oil and grease (FOG) degradation will be less effective than they are for decomposition of more readily degradable organic solids. A common problem in dairy effluent treatment plants is precipitation of fat as a result of lower pH and temperature in the treatment plant. The operation of fat traps is enhanced by ensuring that the effluent is as cool as possible. Milkfat is a liquid at temperatures greater than 35°C and it is difficult to remove in a fat trap. More fat is removed in fat traps at acidic pH values. At pH values above 8.5 the milkfat tends to be emulsified or saponified and therefore is not removable in the fat trap. Dissolved air floatation for fat and suspended solids removal is incorporated in dairy waste treatment to improve efficiency and lower capital costs. Highest concentration of oil and grease was observed in May while lowest value in January. Oil and grease values varied from 2.06±0.06 mg L^{-1} to 5.83±0.10 mg L^{-1} in treated waste water. These values were found within the permissible limits.

Biochemical Oxygen Demand

BOD_5 is the measure of the presence of organic materials in aqueous solution which will be oxidized biologically and in turn will support the growth of micro-organisms (Ciaccio, 1972). The BOD_5 of pure milk comes approximately 100000 mg L^{-1}, thus a BOD_5 coefficient of 1 kg BOD_5 m^{-3} milk is equivalent to 1% of the milk received (Marshall and Harper, 1984). The major constituents which contribute to the BOD_5 of dairy wastes are lactose, milk fat, proteins and lactic acid and the reported average value are 0.65, 0.89, 1.03 and 0.63 kg BOD_5 per kg component, respectively (EPA, 1971). Generally, it has been found from laboratory studies that each kg of milk solids requires about 1.25 kg of oxygen for complete oxidation. Detergents and sanitizers are, potentially, significant contributors to BOD_5, refractory COD and phosphate concentrations in dairy plant. The BOD_5 of milk plant effluent affected by various factors. They depend on the type and qualities of the products manufactured by the plant, the amount of oxygen required for oxidation of different constituents such as fats, carbohydrates and protein are different (Rajsekhar et. al., 2000).

Higher BOD_5 values can also be attributed to lower DO content of the water. Higher BOD_5 values during summer may be associated with the depletion of oxygen at high temperature and increased rate of biochemical reactions which results in increased growth of micro-organisms, thus increasing the rate of microbial decomposition. However, lower BOD_5 values during winter season may be due to increased solubilization of oxygen and decreased rate of decomposition of organic matter at lower temperature resulting in lower consumption of DO. BOD_5 values were found atpar with the prescribed values.

Chemical Oxygen Demand

The chemical oxygen demand of water is the amount of oxygen, required to oxidize the organic material present, by a strong oxidant under reflux conditions. It is considered to be an important parameter because it estimates 95-100% oxidized organic material, while BOD_5 estimates only 70% of oxidisable organic matter. COD values were positively correlated to BOD_5 values.

The ranges of biochemical oxygen demand and chemical oxygen demand concentrations are very large. It should be noted that milk fat is highly biodegerable, as illustrated BOD_5/COD ratio and high BOD_5 rate constant [0.31 (Brown and Pico, 1980)].

Since the COD test is of short analytical duration, it is a useful tool to monitor plant effluent. However, as it does not measure the actual oxygen depletion characteristics of waste water in streams (which can only be dome bio chemically) it is not a direct substitute for the BOD_5 test. The COD test can be used to establish trends, to pinpoint spills promptly and to identify other non-routine problems. The test can be used to monitor BOD_5 test results. Unusual BOD_5/COD ratios may indicate problems such as inadequate BOD seed, poor analytical techniques or equipment malfunctions. Higher values of COD were obtained in summer followed by rainy and winter. This may be associated with the variation in organic matter contents in the dairy waste water. During present investigation or regression model has been developed to predict the BOD values with the help of COD values in the raw and treated dairy waste waters.

Phosphate -P

As phosphate has a limiting or regulating effect on productivity, its estimation is of great importance in determining the biological productivity of aquatic systems. In natural waters, phosphorus occurs as orthophosphate and polyphosphate anions and in trace amounts, as organically bound phosphorus. Results are also supported by the findings of Bhuyan (1970), as he has reported a decrease in phosphate concentration with the increase in dissolved oxygen content of the water of some ancient tanks. Seasonally, higher values of it during summer season may be due to increased decomposition of waste material at higher temperature. Generally, lower values of phosphate during winter indicate its higher precipitation at low pH and temperature.

Phosphate values, which vary independently of BOD_5 values, are in excess amount and offer a potential source of phosphate for enhancing algae growth in natural waterways. Higher concentration of phosphorus in the dairy waste water may be associated with the cleaners used in the dairy plant. Non-phosphate cleaners are not as effective and add to the cost of cleaning because they require higher concentrations, longer cleaning cycles and sometimes the addition of an acid cleaning cycle. Phosphate concentrations vary widely, probably reflecting differences in the amount of phosphate in the cleaning compounds used. High concentrations of acid whey also result in high concentrations of phosphate.

Total Kjeldahl nitrogen as N

Total Kjeldahl nitrogen is sum of ammonia nitrogen and organic nitrogen. This does not include nitrite nitrogen and nitrate nitrogen. The classical Kjeldahl method is used to determine the total nitrogen content. Nitrogen content was recorded maximum in May in case of raw as well as treated waste water. Minimum nitrogen was observed in the month of December.

Milk processing waste contains nitrogen in a variety of forms originally bound nitrogen (primarily as proteins and aminoacids) through ammonium ion (NH_4^+), free ammonia (NH_3) to nitrite $(NO_2)^-$ and nitrate $(NO_3)^-$ ions. A series of processes, including decomposition (of organic nitrogen), nitrification (of ammonia nitrite and nitrate), uptake of soluble forms by crops, soil absorption (of ammonium ion), denitrification (of nitrate to nitrogen gas), and volatilization (of ammonia gas), control the- nitrogen budget of soil. Although nitrogen acts as fertilizer in land application system, infiltration of dissolved nitrate will contaminate ground water. Hence, nitrogen loading is the guiding factor in land application system. Moreover, nitrogen levels in crops can reach levels which are harmful to livestock or are toxic to the crops themselves. Nitrogen values obtained during present investigation were found nearer to the prescribed limits.

Chloride

Chloride in the form of ion (Cl^-) is one of the major inorganic anions found in waste water. The saltiness of water depends to some extent upon the counter cation instead of this anion. A salty taste can be detected in 250 ppm sodium chloride, but not in four times that concentration of chloride ion, if calcium or magnesium are the counter cations (Ciaccio, 1972). The source of chloride content is usually sodium chloride which is used in industry for making products passes unchanged through the drains. Its higher values were observed in summer followed by rainy, and winter in case of raw waste water as well as treated waste water. Higher values observed in summer may be due to higher temperature, similar findings were reported by Rajsekhar et al., 2000.

Sulphate

Sulphate is an important factor, as it produces cathartic effects when present with sodium and magnesium as the counter anions in the water. Low concentration of sulphate is also found responsible for restricting the growth of phytoplanktons (Beauchamp, 1933). Sulphate occur naturally in water as result of leaching from gypsum and other common minerals. In addition, sulphates may be added to water systems in several treatment processes. The sulphate content of municipal water supplies is usually increased during clarification by alum. Sulphates contribute to the total solid content. Sulphate was recorded higher in summer months followed by rainy and winter months. Similar trend was observed in case of raw as well as treated waste water.'

Nutrient removal by aquatic macrophytes

Fresh water aquatic macrophytes grow naturally in water bodies polluted by nutrient loading from urban and agricultural activities. *Eichhornia crassipes* shows luxuriant growth in nutrient rich medium and removes nutrients rapidly. When in combination with small leaf plants, combination with *Lemna minor* and *Azolla pinnata* it gives remarkable results due to synergistic effect. Combination of *Eichhornia crassipes* and *Spirodela polyrhiza* was also recorded as a reliable set for nutrient removal. *Eichhornia crassipes* and *Pistia stratiotes* in combination does not give good results. *E. crassipes* and *P. stratiotes* differ considerably in their pH requirements for optimum growth. *P. stratiotes* has a lesser range of tolerance to variation in hydrogen ion concentration of the substrate than *E. crassipes*. Present findings are similar with the results reported by Chadwick and Obeid, 1966 for other waste water. *P. stratiotes* seems to exhibit an equal plasticity to *E. crassipes* in terms of plant number and an even greater plasticity in terms of individual plant size. Results may now be discussed in relation to 'antagonism' that has been reported between *E. crassipes* and *P. stratiotes* (C.S.A., 1958); similar elimination of the latter by the former was reported by Gay, 1958 in the Nile; the assertion by Parija (1934a) that 'water hyacinth kills Pistia by its luxuriant vegetable growth'.

Highest removal of nitrogen and phosphorus during present investigation confirms the findings of various researchers. Nutrient absorption by controlled population of *E. crassipes* (Mart) Solms (Water hyacinth), or other species of aquatic plants might reduce concentrations of nitrogen and phosphorus in eutrophic lakes or in effluents prior to their release into natural waters (Boyd, 1970; Rogers and Davis, 1972; Steward, 1970; Yount and Crossman, 1970). Steward, 1970 suggested that 1 ha of plants in subtropical regions might remove upto 6,000 kg nitrogen and 600 kg phosphorus per year. Boyd, 1969 using more conservative values for growth and nutrient content, estimated potential annual uptake of nitrogen and phosphorus by water hyacinths in warm climates as 1,980 and 322 kg ha-1, respectively. Rogers and Davis (1972) after measuring nutrient uptake by *E. crassipes* in growth chamber experiments, concluded that absorption by 1 ha of water hyacinths would exceed 2,500 kg of nitrogen and 700 kg of phosphorus per year if maximum growth could be sustained.

Higher absorption of nitrogen in comparison to the phosphorus is similar to the findings of Boyd, 1976, he has reported 5 to 10 times rapid absorption of nitrogen as phosphorus. Nitrogen and phosphorus content of plant tissue was closely associated with the nitrogen and phosphorus content of dairy waste water in the glass aquariums. Uptake of N and P by aquatic macrophytes is directly related to N and P content of dairy waste water. Culley and Epps (1973) found upto 0.8% (8,000 ug/g) of phosphorus in duckweed species collected from sewage lagoons, but they did not relate this concentration of phosphorus to that in the lagoon. However, same workers found up to 2.84% phosphorus in *(Spirodela oligorhiza)* suggesting that this species might be more effective than *Lemna sp.* in accumulating phosphorus from sewage effluent containing >2.1 ug/ml of P.

Dissolved oxygen content of the water under large-leaf floating plants was low. In experimental set containing *Lemna minor* O_2 content of the water was found to be highest. *Eichhornia* forms a dense mesh on the water surface which reduces oxygen diffusion at air-water interface on contrary *Lemna* allows a greater diffusion at air-water interface. Water hyacinths were found to be less efficient in removing phosphorus than nitrogen. Similar findings are reported by Steward, 1970; Dunigan et al., 1975 and Boyd, 1976. Nitrogen was rapidly lost from all system. Nitrogen removal by small leaf floating plants was poor.

In experimental sets containing small leaf floating plants increase in pH created favourable conditions for NH_3 volatilization, rapid nitrification during this study prevented any loss of N through volatilization. The overall nitrogen removal by this system was poor. It should be noted that in our study, plants were cultured in a system containing no underlying sediment, thus reducing N loss due to denitrification. Under natural conditions, NO_3 formed in the water would diffuse into the underlying sediment and undergo denitrification (Engler and Patrick, 1974). Thus, in present study higher removal of nutrients are only due to uptake by macrophytes.

In water hyacinth and water lettuce containing experimental sets, pH values remained relatively constant (6.9-7.3) indicating less favourable conditions for NH_3 volatilization. Under field conditions, dense cover of floating macrophytes (Reddy, 1981) depletes the dissolved oxygen level of the water to <1 mg L^{-1}, creating less favourable conditions for nitrification. Higher absorption of nitrogen by *Eichhornia crassipes* was observed during present study.

The deficient supply of O_2 (by surface aeration) did not affect the purification process. Sculthorpe (1967) has reported that some O_2 produced during photosynthesis is transported to the plant roots whereas Dinges (1982) has reported that atmospheric oxygen enters the stomata of the hyacinth stems/leaves, which is transported down into the roots. However, this oxygen release keeps the micro-organisms on the roots metabolizing aerobically, even though the surrounding water is anaerobic. Lowest concentration of dissolved oxygen in the waste water reveals the fact that due to low oxygen, organic contents are not oxidized. In a study of nitrifier distribution in rivers, Matulewich and Finstein (1978) reported that greater numbers of nitrifiers colonized submerged roots and stems of aquatic plants than other potential attachment sites (rocks, bottom sediments, etc.). In aquatic systems with dense cover of floating plants (water hyacinth or pennywort), denitrification can possibly occur in the root zone of floating plants (Reddy, 1983). Permanent removal of nitrogen from the system can be achieved through denitrification during which nitrate is reduced to nitrite and then to elemental nitrogen, and the nitrogen gas is ultimately exported to the atmosphere. Although denitrification is an anoxic process and may not prevail in aerobic waterbody. The observed strong correlation between the nitrogen removal efficiency and dissolved oxygen level also support to hypothesis as oxygen is essential to nitrification.

It should be noted that nitrification does not occur commonly in facultative ponds due to the low density of nitrifying bacteria in the aerobic zone of the waste stabilization ponds (Ferrara and Avci, 1982). Submerged macrophytes release oxygen as a result of imbalance between photosynthesis and respiration (Bouldin et al., 1974).

Higher removal of nutrients from the waste water by *Lemna minor* may be due to its luxuriant growth. Fast growth rate of *Lemna minor* was also reported by Harvey and Fox, 1973. They found that *Lemna minor L.* would double in frond number every four days when grown in sewage effluent under controlled conditions. They suggested that duckweed could be harvested every four days from a population of plants grown on sewage effluent.

The average initial concentration of P in dairy waste water used in the experiment was 3.7 mg L^{-1}. Considerable variability in the phosphorus content of dairy waste water from day to day was observed. Phosphorus content of aquatic macrophytes was closely associated with the phosphorus content of dairy waste water in the glass aquariums. Loss of phosphorus in control (with no macrophytes) experimental set was probably due to precipitation with Ca compounds at high pH levels. In small leaf plants and control experimental sets, pH of the water reached a maximum 8.9 in the mid afternoon, which probably resulted in the precipitation of P with Ca compounds, and the precipitate thus formed probably deposited on the bottom surface. Similar findings were also reported by Reddy, 1982, 1983.

Although plant uptake played a significant role in the removal of N and P, it did not account for all of the N and P loss from the experimental sets, indicating the possibility of biochemical (denitrification and microbial assimilation) and physico-chemical (volatilization of NH_3) processes functioning in the system. Nitrogen loss due to denitrification probably would be of greater magnitude if anaerobic sediment was present in the experimental sets. Phosphorus loss appears to be due to chemical precipitation with Ca and Mg in control experimental sets. The significance of these processes needs further investigation.

Use of water hyacinth would be limited in tropical and semitropical regions unless protected from freezing temperatures. In temperate regions thermal effluent or some other form of heating would be necessary during the cooler seasons of the year. Culley and Epps (1973) suggested that duckweeds might be a desirable plant for removing nutrients from sewage effluent. In general duckweed species are more tolerant of cool weather than water hyacinth.

Presence of aquatic plants in the overlying waler alters the physico-chemical characteristics (pH, dissolved oxygen, alkalinity, and temperature) of the water (Reddy, 1981). Change in these characteristics can influence the behaviour of nitrogen and phosphorus in the water (Bouldin et al., 1974; Mikkelsen et al., 1978; Reddy and Graetz, 1981). Dissolved oxygen of the water is influenced by the type of aquatic macrophyte. Large leaf plant water hyacinth densely covers the water surface thus, dissolved oxygen is depleted.

Some of the N transformations that can be influenced by the changes in pH, alkalinity, DO, and temperature are NH_3 volatilization, nitrification and denitrification. Increased pH as a result of increased rate of photosynthesis over respiration can enhance NH_3 volatilization losses (Bouldin et al., 1974; Mikkelsen etal., 1978).

Diel variations in pH can also alter the soluble P concentration of the water. Increased pH levels as a result of CO_2 depletion results in precipitation of soluble P with Ca compounds (Reddy and Graetz, 1981) in high alkalinity water. Low DO in water hyacinth ponds can favour the NO_3-N loss through denitrification (Reddy et al., 1980) and increase the availability of soluble P. According to Charles and Claude (1975) and Rai and Munshi (1979) high value of pH were noted in the pond with 0% water hyacinth coverage.

Water temperature is usually lower than that of air temperature. Lower temperature of water below the aquatic macrophyte mat was the direct result of shading and the restriction of mixing of the water by the floating vegetation (Attionu, 1976).

Results presented in this study show that aquatic macrophytes can be effectively used in reducing the N and P levels of nutrient enriched waters. Water hyacinth showed the highest removal rate for both the nutrients. N removal by small leaf floating plants and submerged macrophyte are low and these plants are not suitable for monoculture for waste water treatment. However, these plants can be effectively cultured in combinations along with large-leaf floating plants to enhance the nutrient removal capacity and to improve the physico-chemical environment.

Aquatic weeds are the important sources for biogas generation. Water hyacinth has a capacity to double itself within a short period of time. For effective utilization, hyacinth is cut into small pieces and fed into the biogas plant digester alongwith some water through an inlet pipe. Water and hyacinth are thoroughly mixed in a stirrer and the digested material is withdrawn from an outlet pipe. This slurry is effective when used as manure and the gas produced is good as fuel for cooking and power generation. Gas obtained from a mixture of *Salvinia* weeds and cow dung double the amount from that obtained by cow dung alone (Sharma, 2000).

Many aquatic plants including water hyacinth, have a high protein content (Boyd, 1968, 1969). Plants grown to reduce nutrients from enriched waste water or natural waters could probably be used as animal feed. Aquatic plants have potential as feed stuffs in certain nations, but the economics of harvesting and processing would prohibit their direct utilization as a forage (Boyd, 1969). The water hyacinth, *Lemna* and *Azolla* were appeared to be potential, non-toxic, food sources for animals. The aquatic plants can be harvested easily from the constructed water body. These plants may be powdered and mixed with other animal fodder crop plants to use as animal food.

Thus, the controlled growth of these aquatic macrophytes may provide energy conserving and economical device for the dairy waste water treatment besides the production of animal food materials and biogas as a byproduct.

CHAPTER – VIII
SUMMARY, CONCLUSIONS AND RECOMMENDATIONS

Water is a critical, limited and renewable resource. Waste water released by human activities lead to eutrophication of water bodies. Two important nutrients that can cause eutrophication of water bodies are phosphorus and nitrogen, both of which are released from a variety of sources (industrial and domestic). Phosphorus is required by all living things, become available in large concentrations in treated and untreated industrial and sewage water that enter into the natural aquatic bodies, causes eutrophication and help the enormous growth of green and blue- green algal species. Sometimes algal blooms are formed which become so thick that those at the top shaded those at the bottom; not receiving enough light, those at bottom died. The dead algal debris become food for other micro-organisms, which further help in increasing the population of other micro-organisms. Thus, huge population of microbes consume more and more of the oxygen dissolved in the water. The supply of oxygen in the water soon depleted. Due to low dissolved oxygen content of water, fish and other aquatic organisms die. Because fish require a higher concentration of oxygen in water (Botkin and Keller, 1995).

The eutrophication and pollution problems of an aquatic ecosystem is a challenge to the ecologists and environmentalists, as these are the sources of water supply for drinking, bathing and other domestic purpose. In the past, most vascular aquatic plants growing in such aquatic bodies were considered as a nuisance. However, recently there is growing interest in the potential use of such plants for the treatments of domestic sewage, industrial effluent and aquatic ecosystems. Numerous uses of aquatic plants and their ecobehaviour have been tested and reported by earlier researchers. However, there is paucity of data on the dairy waste water treatment by aquatic macrophytes. Present research work is based on the use of aquatic macrophytes grown individually and in combinations for further removal of nutrients from the dairy waste water discharged after the conventional treatment which carry higher amount of nutrients.

Therefore, present research work was designed to analyse physico-chemical properties of raw and treated waste water of Ramnagar dairy, Varanasi and to evolve effective and economic biological treatment method for further removal of nutrients present in the treated waste water.

At the outset a detailed survey was conducted pertaining to various sources of pollutants, total milk collection, processing and conventional treatment plant of Ramnagar dairy. In order to examine the physico-chemical properties such as temperature, pH, total dissolved solids, total suspended solids, oil and grease, biochemical oxygen demand, chemical oxygen demand, phosphate, total Kjeldahl nitrogen as N, chloride and sulphate were analysed for a period of two years i.e. from August 1998 to July 2000 at fortnight intervals.

Five aquatic macrophytes such as *Eichhornia crassipes, Pistia stratiotes, Lemna minor, Azolla pinnata* and *Spirodela polyrhiza* were selected for the present investigation. Plants were collected from Agrofarm pond of the Banaras Hindu University, Varanasi. Three replicates of each were maintained in 150 Litre glass aquarium. In order to evaluate the actual performance of these selected plants one control set was also taken which was filled only with the waste water with no macrophytes. All the five plants were cultured individually i.e. 100% coverage and in combinations with 50% + 50% coverage of the total surface area of aquarium used for aquaculture. Treated waste water by the conventional treatment plant of Ramnagar dairy was used as a source of waste water.

In order to test the significance and validity of data some statistical calculations were also made. Efforts are also made to establish correlation among various parameters.

General survey of Ramnagar dairy pertaining to milk procurement, processing, sources of pollutants, physico-chemical examinations of raw and treated waste water and aquaculture experiments conducted to evaluate efficiency of some aquatic macrophytes for removal of nutrients from the dairy waste water reveals the following important observations :

Milk procurement and waste water generation

- Maximum milk was procured in Ramnagar dairy in the month of December (1999) as 83278.40 Litre/day and minimum 15706.03 Litre/day in August (1998).
- Waste water generation was recorded highest 333113.60 Litre/day in December (1999) and lowest 50118.60 Litre/day in August (1998).

Waste water characterisation

The liquid waste originates from manufacturing process, utilities and service sections. The individual waste streams from various sources have distinct characteristics. The flow, type and concentration of pollutants vary widely depending upon process, water use and cleaning system. Generally, the total volume of waste water generation per unit of milk processed varies from unit to unit. However, an attempt has been made to have an average value of water consumption per unit of milk processed. The raw water requirement for various purposes in milk processing units can be classified as follows (a) Utilities such as cooling tower, boiler, water softener, back washing, and air compressor cooling tower; (b) Process water; (c) Cleaning-in process (CIP) water; (d) Service water; and (e) Sanitary water.

Physico-chemical properties of dairy waste water

- Temperature values for raw waste water (influent) of dairy industry was recorded minimum 21.30±0.2°C in December and maximum 27.1 1±0.40°C in May. In case of treated waste water (effluent) minimum value of temperature was recorded 19.0±0.08°C in December and maximum 25.3±0.5 °C in the month of May.

- pH of raw waste water ranged from 6.66±0.11 in October to 7.3 1±0.10 in June whereas treated waste water showed 6.68±0.11 in October to 8.45±0.06 in June.

- Total dissolved solids were found to be maximum 1929.04±6.50 mg L^{-1} in May in case of raw waste water and minimum 1674.46±24.04 mg L^{-1} in December. TDSs were recorded maximum 1643.36±12.12 mg L^{-1} in May and minimum 1189.28±5.0 mg L^{-1} in December in treated waste water.

- The highest Total suspended solids values 423.23±11.2 mg L^{-1} were recorded in May and lowest 275.72±4.4 mg L^{-1} were recorded in December in case of raw waste water. Treated waste water showed with highest TSSs Values 95.2 1±3.33 mg L^{-1} in May and lowest 60.75±0.60 mg L^{-1} in December.

- Oil and grease was recorded to be maximum 289.16±4.26 mg L^{-1} in May and minimum 194.30±3.68 mg L^{-1} in January in raw waste water whereas treated waste water showed maximum 5.83±0.10 mg L^{-1} in May and minimum 2.06±0.06 mg L^{-1} in January.

- Biochemical oxygen demand was observed to be maximum 1246.48±13.4 mg L^{-1} in May and minimum 1045.94±29.6 mg L^{-1} in January, however, maximum 39.2 1±0.71 mg L^{-1} in May and minimum 28.05±0.76 mg L^{-1} in the month of January in case of treated waste water.

- Chemical oxygen demand was observed maximum 2245.30±5.16 mg L^{-1} in May in raw waste water while its minimum value 1977.0±4.81 mg L^{-1} in January. In treated waste water COD was observed maximum 235.65±3.25 mg L^{-1} in May and minimum 192.13±1.09 mg L^{-1} in the month of January.

- Phosphorus was recorded to be maximum 7.32±0.03 mg L^{-1} in May and minimum 6.37±0.47 mg L^{-1} in December in raw waste water. Maximum 4.6±0.08 mg L^{-1} in May and minimum 3.08±0.47 mg L^{-1} P was noted in treated waste water in the month of December.

- Nitrogen was recorded minimum 64.0±1.03 mg L^{-1} in January and maximum 77.15±2.65 mg L^{-1} in May in raw waste water. In case of treated waste water N was recorded highest 50.09±0.07 mg L^{-1} in May and lowest 40.05±0.04 mg L^{-1} in January.

- Chloride was observed maximum 144.11±3.70 mg L^{-1} in May and minimum 7 8.0±2.40 mg L^{-1} in the month of December in raw waste water. Treated waste water showed maximum chloride 118.66±0.09 mg L^{-1} in May and 68.45±3.20 mg L^{-1} in December.

- Sulphate was noted to be highest 142.25±1.48 mg L^{-1} in case of raw waste water and lowest 92.54±0.73 mg L^{-1} in May and December month respectively. In case of treated waste water maximum was observed 117.0±1.3 mg L^{-1} in May and minimum 71.28±1.0 mg L^{-1} in December.

From the physico-chemical examination of aforesaid parameters it is revealed that higher concentrations of all the parameters were recorded during summer followed by rainy and winter season.

- Cubic regression model was found best suitable model for the prediction of BOD in the case of dairy waste water.

For raw waste water

$$BOD = -431463.666 + 606.04COD - 0.2829COD^2 + 0.000044COD^3$$

$R=0.87$

For treated waste water:

$$BOD\ 852.30 - 10.59COD + 0.044COD^2 - 0.00058COD^3$$

$R=0.93$

Nutrient removal by selected aquatic macrophytes

- *Eichhornia crassipes* was observed to be most efficient plant for removing nitrogen and phosphorus (71.75±0.20% N and 63.21±0.98% P) from dairy waste water when grown individually followed by *Pistia stratiotes* (64.70±0.40% N and 60.84±0.86% P), *Lemna minor* (62.49±0.39% N and 58.80±0.57% P), *Azolla pinnata* (60.13±0.48% N and 56.27±0.51% P). Least removal was recorded in case of *Spirodela polyrhiza* (57.89±0.04% N and 53.32±1.23% P).
- In combination experimental sets, *E. crassipes* and *L. minor* removed maximum (78.85±0.18% N and 69.37±0.11% P) followed by *E. crassipes* and *A. pinnata* (74.12±0.11% N and 68.74±1.00% P).

- Combinations of morphologically similar plants (large leaf and large leaf, and small leaf and small leaf) were not observed as better combination for instance, *E. crassipes* and *P. stratiotes*, and *L. minor* and *S. polyrhiza*.

- Large leaf plant in combinations with small leaf plant (*E. crassipes* and *L. minor*, and *E. crassipes* and *A. pinnata*) were recorded best possible combinations for nutrient removal from dairy waste water.

- Increase in nitrogen, phosphorus, calcium, potassium, magnesium and crude protein in the tissues of aquatic macrophytes grown in dairy waste water reveals that nutrients from the dairy waste water were removed through uptake.

- Higher concentration of calcium, potassium, magnesium, nitrogen and phosphorus in plant tissues after harvesting (at the end of retention period) indicated its importance as a source of animal feed when compared with the nutrient level of other animal food.

Conclusions

- In Ramnagar dairy about 0.10 MLD milk is processed every day for the production of market milk, ghee, paneer, butter and milk cake, etc.

- Waste water is mostly generated from various processing sections in the dairy and contains wash water, portions of spilled milk, spoiled milk and whey.

- Dairy waste do not contain significant quantities of toxic compounds, although accidental spillage of cleaning and sanitizing chemicals may constitute a hazard.

- During milk processing 0.33-0.35 MLD waste water is generated.

- Raw waste water contains very high BOD, COD, TDSs, Nitrogen and Phosphorus.
- Conventional treatment of Ramnagar dairy minimizes 95% BOD, 89% COD and 98% oil & grease.
- Removal of TDSs, sulphate, chloride, nitrogen and phosphorus was accounted only 17%, 17%, 18%, 35% and 37% respectively which was not satisfactory.
- Higher concentrations of BOD, COD, TDSs, TSSs, nitrogen and phosphorus were recorded during summer followed by rainy and winter season.
- During aquaculture experiments *Eichhornia crassipes* and *Lemna* minor was found best possible combination for removal of nitrogen and phosphorus from the dairy waste water. With the help of *E. crassipes* and *L. minor* about 78.85±0.18% N and 69.37±0.11% P may be removed.
- Combinations of large leaf and small leaf aquatic macrophyte removed higher nitrogen and phosphorus.
- *E. crassipes* and *L. minor* combination may be used for the treatment of dairy waste water as tertiary treatment.
- Higher concentration of nitrogen, phosphorus, sodium, potassium, calcium and magnesium were recorded in the harvested plant tissues of *Eichhornia crassipes* and *Lemna minor.*
- Due to higher concentration of nutrients, *E. crassipes* and *L. minor* plants may be used as a source of animal feed.
- After harvesting water hyacinth plants may be used for the production of biogas, rough quality papers and insulators etc., where as both the plants may be used for production of manure.

Recommendations

For the ECO-FRIENDLY management of the dairy waste water following recommendations are made:

- Due to highest nutrient removal capacity of *E. crassipes* and *L. minor* this combination is recommended for the treatment of dairy waste water as tertiary treatment.
- Due to higher concentration of nutrients in *E. crassipes* and *L. minor,* these plants may be used as a source of animal feed.
- After harvesting water hyacinth plants may be used for the production of biogas, rough quality papers and insulators etc.
- Both the plants may be used for production of manure.
- Treated dairy waste water may also be used for floor washing, gardening and aquaculture.

APPENDICES

Appendix-A

ANOVA of different physico-chemical parameter of raw waste water and treated waste water within months

Parameters	*Raw waste water*		*Treated waste water*	
	df	*F ratio*	*df*	*F ratio*
Temperature	11	101.821	11	267.357
pH	11	20.336	11	218.877
TDSs	11	199.518	11	469.490
TSSs	11	452.431	11	363.283
OG	11	177.774	11	822.130
BOD	11	34.400	11	94.645
COD	11	219.044	11	132.094
Phosphorus	11	5.759	11	19.281
Nitrogen	11	43.308	11	93.188
Chloride	11	893.542	11	556.675
Sulphate	11	302.360	11	945.091

All values are significant at $p < 0.001$.

Appendix-B

Correlation between different parameters of raw waste water, treated waste water (1) and between raw and treated waste water

	temp	ph	tds	tss	og	bod	cod	p	tkn	cl	s	templ	phl
temp	1.00000												
ph	.46002	1.00000											
tds	.94003	.45096	1.00000										
tss	.96092	.57495	.93348	1.00000									
og	.94058	.51969	.95195	.95647	1.00000								
bod	.84457	.32585	.88050	.83717	.89312	1.00000							
cod	.79845	.55161	.73200	.84845	.84469	.74068	1.00000						
p	.59671	.54694	.66985	.68974	.71403	.72753	.58769	1.00000					
tkn	.84609	.35841	.87233	.83686	.84775	.71617	.67950	.52401	1.00000				
cl	.91275	.47509	.89023	.91468	.90432	.81329	.80356	.63285	.72961	1.00000			
s	.93098	.51130	.89768	.92584	.91712	.81719	.81237	.60322	.76528	.95691	1.00000		
templ	.95067	.55806	.95251	.97552	.96418	.85814	.84461	.68422	.86474	.90097	.92694	1.00000	
phl	.56596	.75609	.52213	.65796	.64256	.39820	.71556	.46917	.47644	.46821	.55221	.65347	1.00000
tdsl	.93813	.46639	.98522	.94410	.95210	.88150	.73576	.67792	.87944	.89404	.90610	.96576	.52687
tssl	.94370	.55583	.89808	.96892	.92243	.81967	.86065	.65247	.83912	.90536	.91574	.95541	.60277
ogl	.90279	.55280	.88230	.93502	.93561	.83260	.94565	.65722	.81755	.90347	.89955	.94566	.65171
bodl	.84179	.50240	.77998	.86492	.87883	.81595	.96161	.66741	.70783	.84463	.84847	.87285	.63444
codl	.90225	.52051	.90607	.92362	.91838	.79683	.81249	.60682	.90226	.86456	.88677	.94055	.56465
pl	.80082	.48897	.82937	.85382	.87858	.91679	.73187	.88203	.67641	.79438	.77150	.83938	.51201
tknl	.84510	.39060	.87148	.84646	.85533	.72796	.67859	.49479	.90564	.79764	.79602	.86411	.45592
cll	.88105	.50149	.83878	.89431	.88882	.79261	.85430	.63059	.67948	.96479	.92829	.87078	.52898
sl	.91797	.51770	.86774	.90616	.89179	.78535	.80194	.59113	.73835	.94729	.96925	.89706	.55509

	tds1	tss1	og1	bod1	cod1	p1	tkn1	cl1	s1
tds1	1.00000								
tss1	.92249	1.00000							
og1	.89121	.94540	1.00000						
bod1	.78312	.87064	.95317	1.00000					
cod1	.92225	.93850	.92706	.84037	1.00000				
p1	.84099	.81867	.80699	.79963	.76009	1.00000			
tkn1	.89769	.86121	.83051	.69899	.92488	.67597	1.00000		
cl1	.83429	.87327	.89519	.86056	.80334	.79446	.71702	1.00000	
s1	.87552	.90952	.89112	.84671	.88043	.75276	.80351	.89163	1.00000

All values are significant at $p<0.001$.

Appendix-C

ANOVA of different incubation period and experimental sets in aquaculture study

Nitrogen Removal

Plants	*df*	*Incubation period*	*df*	*Experimental sets*
E	7	226404.737***	2	0.470NS
P	7	65748.624***	2	0.823NS
L	7	76789.618***	2	1.076NS
A	7	43412.214***	2	0.822NS
S	7	63268.984***	2	0.996NS
E+L	7	395985.122***	2	0.574NS
E+A	7	811092.617***	2	1.634NS
E+P	7	122364.699***	2	0.679NS
E+S	7	180278.073***	2	0.127NS
L+A	7	208335.202***	2	1.399NS
L+S	7	164171.821***	2	1.976NS
P+A	7	240401.261***	2	3.Q71NS
P+S	7	1.2274E+06***	2	5.497NS
P+L	7	1.2208E+06***	2	5.743NS
A±S	7	2323.764***	2	0939NS
Control	7	77524.758***	2	2.562NS

**** $p<0.001$, NS = Non significant

Phosphorus Removal

Plants	*df*	*Incubation period*	*df*	*Experimental sets*
E	7	738.11***	2	2.423NS
P	7	1379.193***	2	5.543 NS
L	7	6230.336***	2	7.539NS
A	7	17241.514***	2	11.569NS
S	7	6493.260***	2	4.678NS
E+L	7	196767.030***	2	0.540 NS
E+A	7	8618.125***	2	0.087NS
E+P	7	3182.559***	2	1.908NS
E+S	7	14537.806***	2	0.283NS
L+A	7	11906.746***	2	0.643NS
L+S	7	146233.708***	2	0.334NS
P+A	7	5966.998***	2	0.559NS
P+S	7	457197.976***	2	6.992**
P+L	7	10921.491***	2	2.250NS
A+S	7	8955.903***	2	0.967NS
Control	7	23611.316***	2	0.363NS

*** $p<0.001$, ** p<0.0I, NS =Non significant

E = Eichhornia crassipes, P = Pistia stratiotes, L = Lemna minor, A = Azolla pinnata,

S = Spirodela polyrhiza

REFERENCES

Abdrabo, F.H., Ahmad, N.S., Dawood, A.B.A. and Hassan, F.A.M.1989. Organochlorine, organophosphorus and carbonate pesticides cause changes in some properties of buffaloes milk. Egyptian J. Dairy Sci. (17) 105.

Abu-Hilal, 1987. Distribution of trace elements in nearshore surface sediments from the Jordon Gulf of Aqaba Red Sea. Seagrass Coral organism shell pollution sewage effluent fertilizer industry organic matter mineral. Mar. Pollut. Bull. 18 (4) 190-193.

Adamse A.D. 1966. Bacteriological Studies on dairy waste activate sludge, Meded. Lands Hogesch. Wageningen. 66 (6) 1-80.

Adamse, A.D. 1968. Formation and final composition of the bacterial flora of a dairy waste activated sludge. Water Res. (2) 665-671.

Agami M., Litav M. and Waisel Y. 1976. The effects of various components of water pollution on the behaviour of some aquatic macrophytes of the coastal rivers of Israel. Aquat. Bot. (2) 203-213.

Alexander, M. 1977. Introduction to soil microbiology 2 Ed. John Wiley and Sons, Inc., New York.

Alich, J. and R. Inman. 1976. Energy from agriculture. Proceedings of clean fuels from biomass, sewage, urban refuse, and agriculture wastes. (150) 172.

Allen, S.E. 1974. Chemical analysis of Ecological materials. Blackwell Scientific Publications, Oxford.

Allen, S.E., Grimshaw, H.M. and Rowdand, A.P. 1986. Chemical analysis methods in plant ecology (2nd ed.). Edited by P.B.

Moore and S.V. Chapman. Blackwell Scientific Publications, Oxford, London.

Allenby, K.G. 1981. Some analysis of aquatic plant and their waters. Hydrobiologia. (77) 177-189.

Allsopp, W.H.L. 1960. The manatee ecology and use for weed control. Nature, Lond. (188) 762.

Almer, B., Dickson, W., Eckstrom, C., and Hornstrom, B. 1978. Sulfur pollution and the aquatic ecosystem. Sulfur in the Environment Part II, Ecological Impacts, John Wiley and Sons, New York. pp. 271-311.

An Foras Taluntais. 1974. Methods of Treatment of Milk Processing Wastes. An Foras Taluntais-Dublin.

Anon. 1951. The water hyacinth problem and fish farming. Sci. Cult. (11)182.

Antonie, R.L. and F.M. Welch. 1969. Preliminary results of a novel biological process for treating daily waste, 24th Purdue industrial waste conference. Purdue University.

Antonie, R.L. and Hynek, R.J. 1973. Operating experience with Bio-Surf process treatment of food processing wastes. Proc. 28th Purdue Ind. Waste Conf., Ann Arbor Science, Ann Arbor, Michigan. 849-860.

Aowal, A.F.S.A. and Singh J. 1981. Water hyacinth for treating dairy waste. Proc. Environ. Guidelines for Selected Projects. 16-18 Oct., pp.1-5.

Aoyama, I., Hisao, N. and Ma, S.Y. 1986. Uptake of nitrogen and phosphate, and water purification capacity of water hyacinth. Ber. Ohara Inst. Landw. Biol., Okayama University.(19) 77-89.

Aoyama, I., Nishazaki, H. and Yagi, M. 1987. Uptake of nitrogen and phosphate and water purification capacity by water hyacinth (*E. crassipes* [Mart] Solms.). Soils and Fertilizers. 50 (3) 345-349.

Arbuckle, W.S. 1970. Disposal of dairy wastes in byproducts from Milk, Webb, B.H. and Whittier, E.D. (Eds.) 2nd Edn. AVI Publishing, Westport, Connecticut.

Arbuckle, W.S. 1972. Ice Cream, 2nd edn., AVI Publishing, Westport, Connecticut.

Arceivala, S.J. 1986. Waste water treatment for pollution control. Tata McGraw-Hill, New Delhi.

Archer, A.J. 1985. Disinfection Sunlight photooxidation of organic pollutants in waste water. J. Water Science and Technology. 17 (4-5) 623.

Arne Henrikson and David F. Brakke. 1988. Sulphate deposition to surface waters. Estimating critical loads for Norway and the Eastern United States. Environmental Science and Technology. 22(1) 8-14.

Arnott, H.J. 1966. Studies of calcification in plants. In H. Fleisch, H.J.Blackwood, and M. Owen (ed.), European symposium on calcified tissues. Springer Verlga, Berlin.

Aruw, V., et al. 1988. Biological mechanism of acetate uptake mediated by carbohydrate consumption in excess phosphorus removal systems. Water Research. 22(5) 565.

Attionu, R.H. 1976. Some effects of water lettuce (*P. atiotes L.*) on its habitats. Hydrobiologia 50(3) 245-254.

Ault, T., Velzeboer, R., Zammit, R. 2000. Influence of nutrient availability on phytoplankton growth and community structure in the Port Adelaide River, Australia Bioassay assessment of potential nutrient limitation. Hydrobiologia. 429 (1-3) 89.

Avualt. Jr, J.W. 1980. Aquaculture in fisheries management. ed. R.T. Lackey and L.A. Nielson, Blackwell. 379-411.

Bagnall, L.O., J.A. Baldwin and J.F. Hentges. 1974. Processing and storage of water hyacinth silage. Hyacinth Control J. (12) 73-79.

Bailey, R.C. and Stokes, P.M. 1984. Evaluation of filamentous algae as biomonitors of metal accumulation in software lakes: A multivariate approach, Aquatic Toxicology and Hazard Assessment : Seventh Symposium, ASTM STM 854, American Society for Testing and Materials, Philadelphia. 1985. pp. 5-26.

Baker, R.A. 1961. Problems of tastes and odours. J. Water Poll. Cont. Fed. (33) 1099.

Baltjes J. 1978. Modern waste water control in the dairy industry. Proc. IDF Seminar on Dairy Effluents. Doc. 104, IDF, Brussels. 28-39.

Bandy, J.T. 1984. Water characteristics. J. Water Pollut. Control Fed. 56 (6) 544-548.

Bansho, K. and Miyazaki, A. 1983. Analysis of waste water and effluents. Bunsek. (11) 862-868.

Barchardt, J.A., and H.A. Azad. 1968. Biological extraction of nutrients. J. WPCF. 40(10) 1739.

Bargman, R.D., et al. 1971. Nitrogen-phosphate relationships and removals obtained by treatment processes at the Hyperion Treatment Plant. Pergamon Press Ltd.

Barnett, J.W., Parkin, M.F. and Marshall, K.R. 1982. The characteristics and oxygen demand of New Zealand dairy food plant effluent discharges. Proc. Aquatic Oxygen Seminar, Water and Soil Misc. Publ. No.29. National Water and Soil Conservation Organization, Wellington, New Zealand. 49-53.

Barton, L.V. and Hatchkiss, J.E. 1951. Germination of seeds of *Eichhornia crassipes* Solms. Contr. Boyce Thompson Inst. P1. Res. (16) 2 15-20.

Bartone, B.K. 1981. Irrigation reuse of pond effluent in developing countries. Water Sci. Tech. (19) 289-293.

Bartsch, A.F. 1961. Jour. Water Pollut. Contr. Fed. 33 (3) 239-249.

Bates, G.R. and Phipps 1958. Water hyacinth and its control in Southern Rhodesia. First Afr. Weed Control Conf. 1958. p.6.

Bates, R.P. and Hentges, J.F. 1976. Aquatic weeds eradicate or cultivate? Econ. Bot. (30) 39-50.

Beauchamp, R.S.A. 1933. Suiphates in African Inland Waters. Nature Land. (171) 769-771.

Beckett, R.P. & Brown, D.H. 1984. The control of cadmium uptake in the lichen genus *Peltigera*. J. Exp. Bot. (35) 1071-1082.

Bendell Yong L.I., K.E. Bennett, A. Crowe, C.J. Kennedy, A.R. Kermode, M.M, Moore, A.L. Plant; A. Wood. 2000. Ecological characteristics of wetlands receiving as industrial effluent. Ecological Applications. 10 (1) 310.

Berber, A., and Winter. 1984. The influence of extended anaerobic retention time on the performance of phoredox nutrient removal plant. Water, Science Technology. 17(81).

Beveridge, T.J. 1989. Role of cellular design in bacterial metal accumulation and mineralization. Ann. Rev. Microbial. (43) 147-171.

Bhuyan, B.R. 1970. Physico-chemical qualities of the water of some ancient tanks in Sibsapur, Assam. Environ. Hith. (12) 129-134.

Bishai, H.M. 1960. Effect of *Eichhornia crassipes* on the hydrobiology of Nile water. Rep. Hydrobiol. Res. Unit. Univ. Khartoum. (7) 2-3.

Blanchard, P.E. and Lerch, R.N. 2000. Watershed vulnerability to losses of agricultural chemicals Interactions of chemistry, hydrology, and land-use. Environmental Science and Technology. 34(16)3315.

Bode, H., C.F., Seyfried and A. Kraft 1987. High rate denitrification of concentrated nitrate waste water. Activated sludge waste water treatment. J. 13th Biennial Conf. of the International Water Pollution. Water Sci. Technol. 19(1-2) 163-174.

Bombowna, M. 1985. Ecology of some bodies of water in the forest agricultural basin of the river Brynica near the upper silesian industrial region Poland. 2-chemical composition of water and atmospheric precipitation calcium sulfate, calcium carbonate marls dolomite industrial pollution eutrophication Kozlowa

Gora Swierklaniec Park pond lake, Chechlo Naclo. Hydrobiol. 27(4) 433-450.

Bonhomme, M., Rojella, F., Boisseau, G. and Sibony, J. 1990. Enhancing nitrogen removal in activated sludge with fixed biomass. Water Sci. Technol. (G.B.). (22) 127.

Bose, P.K. 1945. The problem of water hyacinth in Bengal. Sci. Cult. (11)167-71.

Bosset, J.O. and Blanc, B. 1978. Determination of the colour of milk and milk products using a tristimulus reflectance photometer (Hunter System). 20th International Dairy Congress, Paris. E. 427-428.

Botkin, D.B. and Keller, E.A. 1995. Environmental Science. John Wiley & Sons, Inc. U.S.A.

Bouldin, D.R., R.L. Johnson, C. Burda and C. Kao. 1974. Losses of inorganic nitrogen from aquatic systems. J. Environ. Qual. (3) 107-114.

Boyd, C.E. 1968. Evaluation of some common aquatic weeds as possible feed stuffs. Hyacinth Control Jour. (7) 26-27.

Boyd, C.E. 1968. Freshwater plants A potential source of protein. Econ. Bot. (22) 359-368.

Boyd, C.E. 1969. Production, mineral nutrient absorption, and biochemical assimilation by *Justicia americana* and *Alternanthera philxesoides* Arch. Hydrobiol. (66) 139-160.

Boyd, C.E. 1969. The nutritive value of three species of water weeds. Economic Botany. (23) 123-127.

Boyd, C.E. 1969. Vascular aquatic plants for mineral nutrient removal from polluted waters. Econ. Bot. (23) 95-103.

Boyd, C.E. 1976. Accumulation of dry matter, nitrogen and phosphorus by cultivated water hyacinths. Econ. Botany (30) 5 1-56.

Boyd, C.E., and D.H. Vickers. 1971. Variation in the elemental content of *Eichhornia crassipes*. Hydrobiologia (38) 409-414.

Boyle, W.C. and Polkowski, L.B. 1973. Alternate methods of treating or pestreating dairy plants wastes, paper presented at USEPA Technology Transfer Program on the Treatment of Dairy Plant Wastes, Madison Wisconsin.

Bradshaw, A.D., Chadwick, M.J., Jowett, D. and Snaydon, R.W. 1964. Experimental investigations into the mineral nutrition of several grass species. IV. Nitrogen level. J. Ecol. (52) 665-76.

Bremner, J.M., 1965. Inorganic forms of nitrogen. In Black, C.A. (ed.) Agronomy 9 Methods of Soil Analysis. pp. 1179-1237. Amer. Soc. Agron., Madison, WI.

Bremner, J.M., and K. Shaw. 1958a. Denitrification in Soil. I. Methods of investigation. J. Agric. Sci. (51) 22-39.

Bremner, J.M., and K. Shaw. 1958b. Denitrification in soil. II Factors affecting denitrification. J. Agric. Sci. (51) 40-52.

Brenzonik, L. Patrick. 1972. Nitrogen Sources and transformation in natural waters. Nutrients in natural waters. A Willay-Int. Publ. New York (8).

Brix, H. 1994. Use of constructed wetlands in water pollution control Historical development, present status and future perspectives. Water Sci. Technol. 30 (8) 209-223.

Brix, H. and H.H. Schierup. 1989. The use of aquatic macrophytes in water pollution control. Ambio. (18) 100-107.

Brower, G.R. 1985. Industrial wastes and water quality. Wat. Poll. Cont. Fed. 57(6) 625.

Brown, D.H. & Beckett, R.P. 1985. Intracellular & extracellular uptake of cadmium by the moss *Rhytidiadeiphus squarrosus*. Ann. Bot. (55) 179-185.

Brown, H.B. and Pico, R.F. 1980. Characterization and treatment of dairy wastes in the municipal treatment system, Proc. 34th Purdue Ind. Waste Conf., Ann Arbor Science, Ann Arbor, Michigan. 326-334.

Bruce, G., Halstead and Jerry, C. Tash. 1982. Unusual diel pHs in water as related to aquatic vegetation. Hydrobiologia. (96) 217-249.

Brye, Bruce A. 1970. Summary of observed nutrient concentrations and nutrient entrapment of TVA reservoirs. pp. 34-51. In TVA activities related to study and control of eutrophication in the Tennessee Valley. National Fertilizer Development Center, Muscle Shoals, Ala.

Budd, K. and G.W. Kerson. 1987. Uptake of phosphate by two cyanaphytes Cation effects and energetics. Canadian Journal of Botany. 65(9) 1901-1907.

Burden, D.G., Malone, R. and Mericas, C.E. 1987. Isolation as a restoration strategy for nutrient reduction in a small hypereutrophic lake. Fish kill phosphorus eutrophication water quality nutrient loading water pollution. Lowsiana, USA Water/Water Res. 2 1(4) 3 85-394.

Buresh, R.J., and W.H. Patrick, Jr. 1978. Nitrate reduction to ammonium in anaerobic soil. Soil Sci. Soc. Am. J. (42) 913-918.

Burford, J.R., and J.M. Bremner. 1975. Relationships between the denitrification capacities of soils and total water soluble and readily decomposable soil organic matter. Soil Biol. Biochem. (7) 3 89-394.

Burgess, J.E. 1965. Some effects of cultural practices on aquatic environments and native fish population Proc. Annu. Conf. SE Assoc. Game Fish Comm. (19) 413-424.

Burton, H. 1984. Reviews of the progress of dairy Science: the bacteriological, chemical biochemical and physical changes that occur in milk at temperature of l00-15°C. J. Dairy Res. (51) 341-363.

Busk, T.A. and Dierberg. 1989a. Effects of nutrient availability on water hyacinth standing crop and detritus deposition. Hydrobiologia. (174) 15 1-9.

Busk, T.A., Reddy, K.R., Hayes, T.D. and Schwegier, B.R. 1989b. Performance of a pilot-scale hyacinth based secondary treatment system. Journal W.P.C.F. (61) 1217-24.

Buxton, B.M., Ziegler, S.J. and Moore, J.A. 1977. Implication of water quality regulations for Minnesota Dairy Processing Plants, Agricultural Experimental Station Bulletin No. 520, University of Minnesota St. Paul.

C.S.A. 1958. Symposium on *Eichhornia crassipes*. Pubis. Cons. Scient. Afr. S. Sahara. 27.

Cain, J.R., D.C. Paschal and C.M. Hayden. 1980. Toxicity and bioaccumulation of cadmium in colonial green algae Scenedesmus obliquus Arch. Env. Contain. Toxicol. (9) 9-16.

Cairns, J. Jr. 1968. Suspended solids standards for the protection of aquatic organisms. Purde Univ. Engineer Bull. (129) 16-27.

Cairns, John. Jr. and Niederlhner, B.R. 1987. Problems associated with selecting the most sensitive species for toxicity testing. Hydrobiologia. (153) 87-94.

Campbell, R.C. 1986. Statistics for biologists. Cambridge University Press.

Carawan, R.E. and Jones, V.A. 1977. Water and waste management educational program for dairy processing, J. Dairy Sci. (60) 1192-1197.

Carawan, R.E., Chambers, J.V. and Zall, R. 1979. Dairy Processing Water and Waste Water Management, Extension Special Report No. AM-18B. North Carolina Agricultural Extension Service, Raleigh, North Carolina.

Carta, F., Alvarez, P., Romero, F. and Pereda, J. 1999. Aerobic purification of dairy wastewater in continuous regime; reactor with support. Process Biochemistry. (34) 613-619. Elsevier.

Carvalieri, B. 1986. Some effects of Hexavalent Chromium in *L. minor* L. Giomale Botanic Italiano. 120 (1-6) 80-8 1.

Castellfort, J. 1978. Solar powered milk pasteurizer. International Council of Societies of Industrial Design (ICSID).

Center, T.D., and N.R. Spencer. 1981. The phenology and growth of water hyacinth (*Eichhornia crassipes* [Mart] Soims) in a eutrophic northcentral Florida lake. Aquatic Bot. (10) 1-32.

Central Pollution Control Board (C.P.C.B.). 1993. COINDS/40/ 1992-93. Delhi, India.

Chadwick, M.J. 1961. Some observation on the ecology of *Eichhornia crassipes* Solms. Rep. Hydrobiol. Res. Unit. Univ. Khartoum. (8) 23-8.

Chadwick, M.J. and Obeid, M. 1963. The response to variations in nitrogen level of some weed species of the Sudan. Weed Res. (3) 230-41.

Chadwick, M.J. and Obeid, M. 1966. A comparative study of the growth of *Eichhornia crassipes* Solms and *Pistia stratiotes* L. in water culture. J. of Ecol. (54) 563-575.

Chandra, P. and S. Sinha, 1992. Assessment of heavy metal (Cu, Cd, Pb, Cr, Mn) uptake by wetland plant Bacopa monnieri from artificially contaminated sediments. Proc. IV INTECOL. Int. Wetland Conf., Columbus, Ohio, USA. p.118.

Chandra, P., Tripathi, R.D., Rai, U.N., Sinha, S., Garg, P. 1993. Biomonitoring and amelioration of nonpoint source pollution in some aquatic bodies. Wat. Sci. Technol. (28) 323-326.

Charles McVea and Claude B. Boyd. 1975. Effects of Water Hyacinth Cover on Water Chemistry Plant Phytoplankton and Fish in Ponds. J. Environ. Qual., (4) 3, 375-378.

Charpentier, S., Gamier, J. & Flaugnatti, R. 1987. Toxicity and bioaccumulation of cadmium in experimental cultures of duckweed *Lemna polyrrhiza* L. Bull. Env. Contain. Toxicol. (38) 1055-1061.

Chevalier, P. and Noue, J.D. 1985. Efficiency of immobilized hyper concentration algae from ammonium and orthophosphate removal from waste water. Biotech. Lett. (7) 395-400.

Chigbo F.E., Smith R.W., Share F.L. 1982. Uptake of arsenic, cadmium, lead and mercury from polluted waters by the water hyacinth *Eichhornia crassipes*. Environ. Pollut. (A): (27) 18 1-193.

Choung, Y.K., S.J. Jean. 2000. Phosphorus removal in domestic waste water using anaerobic fixed beds packed with iron contactors. Water Science and Technology 41 (1)241.

Ciaccio, L. 1972. Water and water pollution Handbook. Mercel Dekker Inc. New York. (3) 801-802.

Claete, T.E., and P.I. Steyn. 1988. The role of acinetobactor as a phosphorus removing agent in activated sludge. Water Research. 22(8).

Clock, R.M. 1968. Removal of nitrogen and phosphorus from secondary sewage treatment effluent. Ph.D. Diss., Univ. of Florida, Gainesville.

Connell, C.H., and D. Vacker. 1967. Parameters of phosphate removal by activated sludge. Proceedings 7 Industrial Water and Waste Conference, University of Texas, Austin, Texas. II: 28-37.

Convention on Wetlands (RAMSAR). 1998. The key role of wetlands in addressing the global water crisis. International conference of water and sustainable development, Paris, Mar. 19-21, 1-4.

Cooke, J.G. 1994. Nutrient transformations in a natural wetland receiving sewage effluent and the implications for waste treatment. Water Sci. Technol. (29) 209-217.

Cooper, J.S. 1974. Research and development within dairy effluent treatment. Dairy Effluent Treatment, Doc. 77, IDF, Brussels. 3-7.

Coruwell, D.A., J. Zoltek, Jr. C.D. Patrinely, T.D. Furman and J.I. Kim. 1977. Nutrient removal by water hyacinths. J. Water Pollut. Centr. Fed. (49) 57-67.

Cossa, C.B. and J. Noel. 1987. Concentration of mercury in near shore surface waters of the bay of biscay and in the Gironde Estuary France Mar. Chem. 20(4) 389-396.

CPCB. 1998. Dairy Industry-Effluent Quality. Delhi, India.

Crisman, T.L., Schuize, R.L., Brezonik, P.L., and Bloom, S.A. 1980. Acid Precipitation the biotic response in Florida lakes, Proc. Int. Conf. Ecol. Impact. Acid Precip. SNSF Project.

Crist, R.H., Oberholser, K., Shank, H. and Nguzen, D. 1981. Nature of bonding between metallic ions and algal cell walls. Env. Sci: Technol. (15) 1212- 1217.

Crowder, A. 1991. Acidification, metals and macrophytes. Env. Pollut. (71) 171-203.

Crumpton, W.G., T.M. Isenhart and C.M. Hersch. 1987. Determination of nitrate in water using ammonia probes and reduction by titanium III, water pollution waste water sampling cost environmental surveillance. J. Water Pollut. Control Fed. 59(10) 905-908. Culley, D.C., Jr. and A.E. Epps. 1973. Use of duckweed for waste treatment and animal feed. J. Water Poll. Control Fed. (45) 337-347.

Cully, D.D. Jr., Rejmankova J., Kvet J. and Frye J.B. 1981. Production, chemical quality and use of duckweeds (Lemnaceae) in aquaculture. Waste Mgmt. Animal Feed. World Mariculture Soc. (12) 27-49.

Cunningham, W.P. and Saigo, B.W. 1995. Environmental Science. WmC. Brown Publishers, U.S.A.

Dale, H.M. and Gilliespie, T.J. 1976. The influence of floating vascular plants on the diurnal fluctuations of temperature near the water surface in early spring. Hydrobiol. (49) 245-256.

Dalsgaard, T. and N.P. Revsbech. 1992. Regulating factors of denitrification in trickling filter biofilms as measured with the oxygen/nitrous oxide microsensor. FEMS Microbiol. Ecol. (101) 15 1-164.

Davelaar, D., et al. 1978. The significance of an anaerobic zone for the biological removal of phosphate from waste waters. Water SA. 4(2) 54.

Davis, L.S., J.P. Hoffmann, and P.W. Cook. 1990. Production and nutrient accumulation by periphyton in a waste water treatment facility. J. Phycol. (26) 617-623.

Dawson, R.N., and K.L. Murphy. 1972. The temperature dependency of biological denitrification. Water Res. (6) 7 1-83.

DeBusk, T.A. 1982. Standing crop changes, detritus production and decomposition of *Eichhornia crassipes* (Mart.) Solms. M.S. Thesis. Florida Institute of Technology. Melbourne. p.93.

DeBusk, T.A. and Dierberg, F.E. 1989. Effect of nutrient availability on water hyacinth standing crop and detritus deposition. Hydrobiologia. (174) 151-159.

DeBusk, T.A., Reddy, K.R., Hayes, T.D. and Schwegier, B.R. Jr. 1989. Performance of a pilot scale hyacinth based secondary treatment system. J. Water Poll. Cont. Fed. (61) 1217-24.

DeBusk. T.A., and F.E. Dierberg. 1984. Effect of nitrogen and fiber content on the decomposition of the water hyacinth (*Eichhornia crassipes* [Mart.] Soims). Hydrobiologia. (118)199-204.

Dierberg, F.E. and P.L. Brezonik. 1981. Nitrogen fixation (acetylene reduction) associated with decaying leaves of pond cypress (Taxodium distichym var. nutans) in natural and sewage-enriched cypress dame. Appi. Environ. Microbiol. (41) 1413-1418.

Dinges, R. 1978. Upgrading stabilization pond effluent by water hyacinth culture. J. Water Poll. Control Fed. (50) 833-845.

Dinges, R. 1982. Natural systems for water pollution control. Van Nostrand Reinhold, New York.

Doedens, H. 1974a. Aerobic biological treatment of dairy waste water- excess sludge and effects of variations in pH. Dairy Effluent Treatment Doc. (77) IDF, Brussels. 108-121.

Duarte, C.M. and Kaiff, J. 1990. Biomass density and the relationship between submerged macrophytes biomass and plant growth. Hydrobiologia 196 (1) 17-24.

Dunigan, E.P., R.A. Phelan, and Z.H. Shamsuddin 1975. The use of water hyacinths to remove nitrogen and phosphorus from eutrophic waters. Hyacinths Control J. (13) 59-62.

Dunn, J.G. 1967. Diurnal fluctuations of physico-chemical conditions in a shallow tropical pond. Limnol. Oceanogr. (12) 15 1-154.

Dwivedi, U.P. 2000. Impact of different pollutants on physico-chemical and biological characteristics of selected ponds. Ph.D. Thesis, Banaras Hindu University, Varanasi, India.

Dye, C.W., Jones, D.A., Ross, L.T. and Gernert, J.L. 1980. Diel variations of selected physicochemical parameters in lake Kissimmee. FL. Hydrobiologia. (71) 5 1-60.

Dymond, G.C. 1949. The water hyacinth a cinderella of the plant world. In Van Vuran, J.P.J. (ed.). Soil fertility and sewage. Dover Publications, New York.

Eckenfelder Wesley W., Jr. 1966. Industrial Water Pollution Control. McGraw-Hill Book Company.

Eckles, C.H., Combs, W.B. and Macy, H. 1973. Milk and milk products. Tata Mc Graw Hill Publ. Company Ltd., New Delhi.

Edmondson, W.J. 1974. Freshwater biology. John Wiley and Sons, Inc., New York, London.

Elfline, G.S. 1987. Methods of removing heavy metal from waste water streams. US Patent Sludge Waste Water Treatment. 1080 (1) 326.

Emori, H., Nakamura, H., Sumino, T., Takeshima, T., Motegi, K., and Tanaka, K. 1994. High rate and compact single sludge pre-denitrification process for retrofit. Water Sci. Technol. (G.B.). (30) 31.

Engler, R.M., and W.H. Patrick, Jr. 1974. Nitrate removal from floodwater overlying flooded soils and sediments. J. Environ. Qual. (3) 409-4 13.

Environment Protection Act. 1986. Sharma, B.K. 2000. Environmental Chemistry. Goel Publishing House, Krishna Prakashan Media (P) Ltd., Meerut, India.

EPA. 1971. Dairy food plant waste & waste treatment practices. Harper, W.J., Blaisdell, J.L. and Grosskopf, J. (eds.). USEPA 12060 .EGU 03/71, USEPA, Washington, D.C.

Eriksson Peder G. and Weisner Stefan E.B. 1997. Nitrogen Removal in a Waste Water Reservoir, The importance of denitrification by Epiphytic Bioflims on Submersed Vegetation. J. Environ. Qual. (26) 905-9 10.

Everard M. and Denny P.A. 1985. Flux of lead in submerged plants and its relevance to fresh water system. Aquat. Bot. (21) 181-193.

Falbo M.B., Weaks T.E. 1990. A Comparison of *Eichhornia crassipes* (Pontedariaceae) and Sphagnum guinguefarium (Sphagnaceae) in treatment of acid mine water. Econ. Bot. (44) 40-49.

Fayed, Semi, E. and Hussaini, A.S. 1985. Accumulation of Cu Zn Cd and Pb by aquatic macrophytes. Environment International. (11) 77-87.

Felfoldy, L. 1972. Biological control of water quality (in Hungarian). Personnel Communication.

Ferrara, R.A. and Avci, C.B. 1982. Jour. Water Pollut. Contr. Fed. 54 (4) 36 1-369.

Ferrier Pages, C., Gattuso, J.P., Dallot, S., Jaubert, J. 2000. Effect of nutrient enrichment on growth and photosynthesis of the zooxanthellate coral Stylophorapistillata. Coral Reefs. 19 (2) 103.

Findlay, D.L. 1984. Effects on phytoplankton biomass, succession and composition in Lake 223 as a result of lowering pH levels from 5.6 to 5.2. Data from 1980-1982, Can. M/s Report of Fish. Aquat. Sci. No. 1761. p.10.

Finstein, M.S. and J.V. Hunter. 1967. Hydrolysis of condensed phosphate during aerobic biological sewage treatment. Water Res. (1) 247-254.

Fjaervoll, A. 1970. Anhydrous milk fat, manufacturing techniques and future applications, Dairy Inds. (35) 424-428.

Flavia F. Coelho, Frederico S. Lopes, Carlos F. Sperber. 2000. Density dependent morphological plasticity in Salvinia auriculata. Aquatic Botany. (66) 273- 280.

Forstner, U. and G.T.W. Wittman, 1983. Metal pollution in Aquatic Environment. Spinger Verlag, Berlin, Heidelberg, New York, Tokyo.

Fortier, J.L., Reboul, B., Philip, P., Simard, M.A., Picker, P., and Jolicoeur, C.1980. Calorimetric studies of biodegradation process in biological waste treatment plant. J. Water Pollution Control Fedn. 52 (1), 89.

Fuchs, G.W. and M. Chen. 1975. Microbial basis of phosphate removal in the activated sludge process for the treatment of waste water. Microbial Ecology. No.119.

Gajghate, D.G. and Reddy P.J. 1989. COD BOD relationships for Industrial Wastes. Indian J. Env. Prot. 9 (11) 805-807.

Gakstatter, J.H., Allum, M.O., Dominguez, S.E. and Crouse, M.R. 1978, Jour. Water Pollut. Control Fed. (50) 718-722.

Galpin, D.B. 1981. Effluent disposal from New Zealand dairy plants. N.Z. J. Dairy Sci. Technol. (16) 289-292.

Ganapati, S.N. and P.1. Chacko 1951. An investigation of the river Godavari and the effects of the paper mill pollution at Rajahmundry. Proc. Indo-Pac. Fish Counc. Madras Meeting Sec :11 and III. 70.

Ganf, G.G. 1974. Diurnal mixing and the vertical distribution of phytoplankton in a shallow equatorial lake (Lake Georg). J. Ecol. (62) 611-629.

Garg, P. and P. Chandra. 1990. Toxicity and accumulation of Chromium in Ceratophyllum demersum L. Bull. Env. Contain. Toxicol. (44) 473-478.

Gasser, J.K.R. 1964. Some factors affecting losses of NH3 from urea and (NH4)2504 applied to soil, J. Soil Sci. (15) 258.

Gaudet, J. 1977. Uptake, accumulation and loss of nutrient by papyrus in tropical swamps. Ecology. (58) 4 15-422.

Gay, P.A. 1958. *Eichhornia crassipes* in the Nile of the Sudan. Nature, Lond. (182) 538-9.

Gay, P.A. and Berry, L. 1959. The water hyacinth a new problem on the Nile. Geogri. J. (125) 89.

Gearheart, R.A. and M. Higley. 1993. Constructed open surface wetlands the water quality benefits and wildlife benefits-city of Arcata, CA. p.56 1-567. In G.A. Moshiri (ed.) constructed wetlands for water quality improvement. Lewis Publ., Boca, Raton, FL.

Geldard, F.A. 1953. The human senses. John Wily and Sons, Inc., New York. pp.270-295.

Gerber, A., Mostert, E.S., Winter, C.T. and Devilliers, P.H. 1987. Interactions between phosphate and organic substrate in biological nutrient removal processes. Activated sludge waste water treatment. Water Sci. Tech. 19 (1-2)183-194.

Gersberg, R.M., B.V. Elkins, S.R. Lyon, and C.R. Goldman. 1986. Role of aquatic plants in waste water treatment by artificial wetlands. Water Res. (20) 363-368.

Gloyna, E.F. 1971. Waste Stabilization Ponds, WI-JO Monograph Series, 60. WHO, Geneva.

God, P.K. and R.K. Trivedi. 1984. Some considerations on sewage disposal to freshwater and resultant effects. Poll. Res. (3) 7-12.

Goldman, C.R. and Wetzel, R.G. 1963. A study of the primary productivity of Clear Lake, Lake County, California. Ecology (44) 283-294.

Golterman, H.L., R.S. Clymo and Ohnsland. 1969. Methods for physical and chemical analysis of freshwaters. Blackwell Scientific Publications Onsey Mead. Oxford.

Golueke, C.G. 1964. Harvesting and processing sewage grown planktonic algae. Sanitary Engineering Research Laboratory, Univ. California, Berkeley, SERL Report No. 64-8, p.55.

Grobbelaur, J.V. 1989. The contribution of phytoplankton productivity in turbid fresh water to their trophic status. Hydrobiologia. 17 (3) 127-133.

Guilizzoni, P. 1991. The role of heavy metals and toxic materials in physiological ecology of submerged macrophytes. Aquatic Bot. 87-109.

Guilizzoni, P.G., Galanti and H. Muntau. 1989. The aquatic macrophytes of lake Maggiore Species composition, spatial distribution and heavy metal concentration in tissue. Mem. 1st Ital. Idrobiol. (46) 23 5-260.

Hach. 1969. Water and waste water analysis. Procedures. Hach Chemical Co., Ames, Iowa. p.105.

Haider, S.Z. 1984. Mechanism of adsorption of chemical species from aqueous medium by water hyacinth and prospects of its utilization. International symposium on water hyacinth, Hyderabad. Proceedings. p.4 1-57.

Haigh, J.C. 1940. The propagation of water hyacinth (*Eichhornia crassipes* Soims.) by seed. Trop. Agric. Mag. Ceylon Agric. Soc. (94) 296-7.

Hall, C.W. and Hedrick, T.I. 1966. Drying Milk and Milk Products, AVI Publishing, Westport, Connecticut.

Hammer, D.A. (ed.) 1989. Constructed wetlands for waste water treatment Municipal, industrial and agricultural. Lewis Pubi., Boca Raton, FL.

Hammer, D.A. and R.L. Knight. 1993. Designing constructed wetlands for nitrogen removal. Water Sci. Technol. 29 (4) 15-27.

Handbook of Animal Husbandry. 1990. Publication and information division. Indian Council of Agricultural Research, PUSA, New Delhi, India.

Hanne Werner. 1978. Methods of Analysis for Dairy Effluents, LDF Bulletin.

Haque, A. and Sharma, S. 1986. Water hyacinth to fight water pollution. Science Reporter. Dec. 757-62.

Harold R. Jones. 1974. Pollution control in the Dairy Industry. Noyes Data Corporation, Park Ridge, New Jersey.

Harper, W.J. and Chambers, J.V. 1978. Upgrading dairy waste treatment systems, Proc. IDE Seminar on Dairy Effluents, Doc. 107, IDF Brussels. 173-175.

Harvey, N., R. Offer, A. Hades, I. Ravina. 1999. Waste water irrigation-economic concerns regarding beneficiary and hazardous effects of nutrients. Water Resources Management. 13(5) 303.

Harvey, R.M. and Fox, J.L. 1973. Nutrient removal using *Lemna minor.* J. Wat. Pollut. Contr. Fed. (45) 1928-1938.

Hasan, A., Satyakala, S.B. and Kaiser, J. 1986. Effect of aquatic pollution on the mineral accumulation in water hyacinth. Proc. Indian Acad. Sci. Plant Science. 1 (76) 37-40.

Hauser, J.R. 1984. Use of water hyacinth aquatic treatment systems for ammonia control and effluent polishing. J. Wat. Poll. Cont. Fed. (55) 2 19-26.

Havens, K.E. 1984. Freshwater plankton response to acidification, Ph.D. Diss. W. Virginia University.

Hellquist, C.B. 1980. Correlation of alkalinity and the pond distribution of *Potamogeton* in New England, Rhodora. (82) 33 1-344.

Hemmings, M.L. 1980. The treatment of dairy wastes, Dairy Inds. Int. 45 (11) 23-28.

Hendry, G.R. 1976. Effecfs of pH on the growth of periphytic algae in artificial stream channels. Sur. Nedbors Virkning pa Skog og Fisk Project IR. 25/76. Hendry, G.R., and Vertucci, F. 1980. Benthic plant communitic in acidic lake Colden, New York *Sphagnum* and the algal mat, Ecological Impacts Acid precipitation, SNSF Project. 314-315.

Hepher, B. 1958. On the dynamics of phosphorus added to fish ponds in Israel. Limnol. Oceanogr. (3) 84-100.

Hill, B.H. 1979. Uptake and release of nutrients by aquatic macrophytes. Aquat. Bot. (7) 87-93.

Hillebrand, H., Sommer U. 2000. Effect of continuous nutrient enrichment on microalgae colonizing hard substrates. Hydrobiologia. 426 (1-3) 185.

Ho, G.E., Gibbs, R.A., Mathew, K. and Parker, W.F. 1992. Ground water recharge of sewage effluent through amended sand. Wat. Res. (26) 285-293.

Ho, G.E., Mathew, K. and Gibbs, R.A. 1992. Nitrogen and phosphorus removal from sewage effluent in amended sand columns. Wat. Res. 26 (3). p.295 300.

Holm, L.G., L.W. Weldon, and R.D. Blackburn. 1969. Aquatic Weeds. Science. (166) 699-709.

Hong, S.N. et al. 1982. A biological waste-water treatment system for nutrient removal. EPA workshop on Biol. Phos. Removal, Annapolis, Maryland.

Hong, S.N., and K.L. Andersen. 1993. Converting a single sludge oxygen activated sludge system for nutrient removal. The 66th WEF Conference, Anaheim, CA.

Horton, J.P. and Trebler, H.A. 1953. Recent developments in the design of small milk waste disposal plants, Can. Dairy Ice Cr. J. 33(4) 38-41; 78.

Hosetti, B.B. and H.S. Patil. 1986. Impact of *Lemna minor* on the effluent quality of sewage stabilization pond. Geobios. 13(6) 244-247.

Hosetti, B.B., H.S. Patil and S.V. Hosmani. 1985. Treatment of dairy animal waste with algae Indian J. Ecol. 12 (1) 8-12.

Howard-Williams, C. 1981. Studies on the ability of *Potamogeton pectinatus* community to remove dissolved nitrogen and phosphorus compounds from lake water. J. Appl. Ecol. (18) 6 19-637.

Howard-Williams, C. 1985. Cycling and retention of nitrogen and phosphorus in wetlands. A theoretical and applied perspective. Freshwater Biol. (15) 391-431.

Howeler, R.H., and D.R. Bouldin. 1971. The diffusion and consumption of oxygen in submerged soils. Soil Sci. Soc. Amer. Proc. (35) 202-208.

Huang, J., R.B. Reneau, C. Hagedorn. 2000. Nitrogen removal in constructed wetlands employed to treat domestic waste water. Water Research. 34 (9) 2582.

Hussainy, S.U. 1979. Prog. Water Technol. 11(4/5) 3 15-337.

Hutchinson, G.E. 1957. A treatise on Limnology, Geography, Physics and Chemistry. John Wiley and Sons, New York, Vol.1. 1015.

Ice, J., R. Couch. 1987. Nutrient absorption by duckweed. J. of Aquatic Plant Manag. (25) 30-31. Indian Council of Medical Research,

India. 1962. Ministry of Health Committee on Public Health Engineering Manual and Code of Practice. Manual of Water Supply, New Delhi.

Indian Standard Institution. 1982. Tolerance limits for inland surface waters subject to pollution (Sec. Rev. Ed.) IS 2296-1982 7-14.

Ingemarsson, B. 1986. Patterns of N uptake, accumulation and assimilation in Lemna gibba, when grown with exponentially increasing daily doses of N. Fund. Ecolo. and Agri. aspects of N2 Metabo. in higher pts. Dordrecht. Netherlands, Martinus Nijhaff Publishers. 65-69.

Jackson, G.A. 1980. Marine biomass production through sea weed aquaculture. In Biochemical and photosynthetic aspects of Energy production, ed, A. San Pietro. p. 31-58. Academic Press, New York.

Jackson, M.B. 1985. Filamentous algae in Ontario Sofiwater lakes. Int. Symp. on Acidic Precipitation, Muskoka Canada. pp.1 17-118.

Jackson, M.L. 1958. Soil Chemical Analysis. Prentice-Hall Inc., Englewood Cliffs, N.J. p.498.

Jackson, M.L. 1962. Soil Chemical Analysis, Inc. Englewood Cliffs N.J. U.S.A. 183- 190.

Jam, S.K., Vasudevan, P. and Jha, N.K. 1990. *Azolla pinnata* and *Lemna minor* for removal of lead and zinc from polluted water. Water Research. 24(2) 177-183.

Jana; S. 1988. Accumulation of Hg and Cr by three aquatic species and subsequent changes in several physiological and biochemical parameter. Walt, Air, Soil Pollut. (38)105-109

Jaya Kumar, R. 1985. Categorisation of some important weeds based on nutrients concentration. Maltras Agri. J. 72 (11) 646-649.

Jepson, F.P. 1933. The water hyacinth problem in Ceylon. Trop. Agric. Mag. Ceylon Agric. Soc. (81) 339-55.

Jha, U.N. 1968. The pond ecosystem, Ph.D. Thesis, Banaras Hindu University, Varanasi.

Jhingran, V.G., Natrajan, A.V., Banerjee, S.M. and David A. 1969. Methodology on Reservoir Fisheries Investigation in India. Bull. Cent. Inl. Fish. Res. Inst. (13) 108-112.

Jodrai, M., Martinez, P., Angnlo, R., Gallrego, M.C. and Pozo, R. 1995. Effect of the pasteurization process on DDT and its metabolites. Alimant. (33) 25-28.

John Stone O. Young. 1975. Seasonal and diurnal changes in the water temperature of a temperate pond (England) and a tropical pond (Kenya). Hydrobiologia. 47 (3-4) 5 13-526.

John, To Kirk. 1985. Effect of suspensoids (Turbidity) on penetration of solar radiation in aquatic ecosystem. Hydrobiologia. (125) 195-208.

Johnson, W.K. and Schroefer, C.J. 1964. Nitrogen removal by nitrification and denitrification. J. Water Pollut. Control Fed. (36) 10 15-36.

Jones, Iwan J., John W. Eaton, Hardwick Keith. 2000. The effect of changing environmental variables in the surrounding water on the physiology of *Elodea nuttalie*. Aquatic Botany. (66) 115-129 Elsevier No.2.

Joshi, M.C., Thukral, A.K. and Chand, R. 1982. Cu, Zn, and Mn concentration in leaves of *Tangustata* growing in tailing pond at Khetrinagar, Rajasthan. Indian J. Environ. Health. 24(3) 234-236.

Kadlec, R.H. 1995. Overview Surface flow constructed wetlands. Water Sci. Technol. 32 (3) 1-12.

Kadono, Y. 1982b. Occurrence of aquatic macrophytes in relation to pH, alkalinity, Ca, Cl and conductivity. Jap. J. Ecol. (32) 39-44.

Kaul, V. 1981. Role of macrophytes in aquatic ecosystems of Kashmir. J. of I.BS. 3(60) 16-20.

Kaul, V. 1985. Primary productivity of inland aquatic ecosystem under varying climate. A review Trop. Ecol.

Kaul, V., Trisal, C.L. and Kaul, S. 1980. Mineral removal potential of some macrophytes in some aquatic lakes of Kashmir. J. Ind. Bot. Soc. (59) 108-118.

Kearney, A.T., Inc. 1973. Development document for effluent limitation guidelines and standards of performance. Dairy Product Industry, USEPA 68-1 1-1502 US Environmental Protection Agency, Washington, DC.

Keddy, P., Gaudet, C., Fraser, L.H. 2000. Effects of low and high nutrients on the competitive hierarchy of 26 shoeline plants. J. of Ecology. 88 (3) 413.

Keeney, M. and Bassette, R. 1959. Detection of intermediate compounds in the early stages of browning reaction in milk products. J. Dairy Sci. (42) 945-960.

Khan, A.A., Siddiqui, A.Q. and Nazir, M. 1970. Diurnal variations in a shallow tropical fresh water fish pond in Shahjahanpur, U.P. (India). Hydrobiologia. (35) 297-303.

Khan, K.R. and R.H. Siddiqui. 1972. Treatment of waste water by biological discs. Indian J. Env. Health. 14.

King, J.E., and A.L. Smith, Jr. 1947. Investigations of the effect of water, hyacinth on the fish and fish habitat of Louisiana waters. USD1, Fish wildlife spec. Sci. Rep.(39) 1-29.

—

Kingsford, M., C.D. Stevenson and W.H.L. Edgerley. 1973. Collaborative test of water analysis (The Chemaqua Programme) Na, K, Ca, Mg, Cl, SO_4, HCO_3, CO_3 and conductivity, N. Z. JI. Sci. (16) 895-902.

Knipling, E.B., S.H. West, and W.T. Hailer. 1970. Growth cha icteristics, yield potential, and nutritive content of water hyacinths. Soil Crop Sci. Soc. of Florida. (30) 5 1-63.

Kozlowoski, T.T. 1986. Measurement of effects of environmental and industrial chemicals on terrestrial plants. Soils of Fertilizers. 49 (1) 49-103.

Krichten, D.J., et al. 1985. Applied biological phosphorus removal technology for municipal waste water treatment by the A/O Process. In Management strategies for phosphorus in the environment. London: Selper Ltd.

Kuehner, R.L. 1964. Recent advances in odour : Theory measurement and control. Ann. N.Y. Acad. Sci. 116 (Art.2) : 3 57-746.

Kumar, P. 1986. Use of water hyacinth in waste water treatment and biogas production. Ph.D. Thesis, University of Roorkee, Roorkee, India.

Kumar, P., and Garde, R.J. 1989. Potentials of water hyacinth for sewage treatment. Res. J. Water Pollution Control Fed. 61(11/12)1702-1706.

Lai, C.C., Patrick and Lam, K.S. Paul. 1997. Major Pathways for nitrogen removal in waste water stabilization ponds. Water, Air and Soil Pollution. (94) 125-136.

Lakshman, G. 1979. An ecosystem approach to the treatment of waste water. J. Environ. Qual. (8) 353-61.

Lavoie, A. and DeLa Noue J. 1983. Harvesting microalgae with chitosan. J. World Maricul. Soc. (14) 685-694.

Lawrence George H.M. 1967. Taxonomy of vascular plants. Mohan Primilani, Oxford and IBH Publishing Co., 66, Janpath, New Delhi and printed at Ram Printograph (India) New Delhi 110020.

Lawrence, J.M. and L.L. Weldon. 1965. Identification of aquatic weeds. Hyacinth Control J. (4) 5-17.

Lazarek, 5. 1983. Structural and functional aspects of epiphytic and benthic algae in the acidified Lake Gardsjon, Sweden, Ph.D. Dissertation, University of Lund. p.112.

Leakovic S., I. Mijatovic, S. Cerjanstefanovic, E. Hodzic. 2000. Nitrogen removal from fertilizer waste water by ion exchange. Water research. 34 (1) p.185.

Levin, G.V. and J. Shapiro. 1965. Metabolic uptake of phosphorus by waste water organisms. J. W.P.C.F. 37 (6) 800.

Levitt, Jacob. 1969. Introduction to plant physiology. C.V. Mosby Co., St. Louis. p.304.

Liaw, W.K., and H.R. MacCrimmon. 1978. Assessing changes in biomass of riverbed periphyton. Int. Rev. Gesamten. Hydrobiol. (63) 155-172.

Lind, C.B. 1998. Phosphorus inactivation in waste water treatment. Biological and chemical strategies. Water/Engineering and Management. pp. 18-21.

Lock, M.A. 1993. Attached microbial communities in rivers. In T.E. Ford (ed.) Aquatic microbiology. Blackwell Scientific, Oxford. pp. 113-138.

Lorenzen K. 1999. Nitrogen recovery from shrimp pond effluent dissolved nitrogen removal has greater overall potential than particulate nitrogen removal, but requires higher rates of water exchange than presently used. Aquaculture Research 30 (11-12) p.923.

Luijn, F., P.C.M. Boers, and L. Lijklema. 1996. Comparison of denitrification rates in lake sediments obtained by the N_2 flux method, the ^{15}N isotope pairing technique and the mass balance approach. Water Res. (30) 893-900.

Lytken, B. 1974. Relations between production size and effluent pollution, Dairy Effluent treatment, Doc. 77 IDF, Brussels. 24-28.

Mackenthum, K.M. 1964. Limnological aspects of recreational lakes. U.S. Gov. Printing Office, Washington, D.C. p.176.

Mackenthum, K.M. 1965. Nitrogen and phosphorus in water an annotated bibliography of their biological effects. U.S. Gov. Printing Office, Washington, D.C. p.111.

Madsen, J.D., and M.S. Adams. 1988. The nutrient dynamics of a submersed macrophyte community in a stream ecosystem dominated by *Potamogeton pectinatus*. L.J. Fresh water Ecol. (4) 54 1-550.

Magnusson, F. 1974. Spray irrigation of dairy effluent. Dairy Effluent Treatment, Doc. 77, IDF, Brussels. 122-130.

Mangi J., Schmidt K., Pankow J., Gaines L., Turner P. 1978. Effect of chromium on some aquatic plants. Environ. Pollut. (16) 285-29 1.

Manja, K.S. 1986. Detection of faecal pollution in drinking water in developing countries a simple and low cost method. Ecology. 1(3) 23-27.

Manny, B.A., Nichols, S.J., and Scholesser, D.W. 1991. Heavy metals in aquatic macrophytes drifting in a larger river. Hydrobiol. (219) 333-344.

Manson, J.G. and Manson, B.E. 1958. Water hyacinth reproduces by seed in New Zealand. N.Z. JI Agric. (96) 191.

Marshall, K.R and Harper, W.J. 1984. The treatment of wastes from dairy industry. New Zealand Dairy Res. Inst.

Marshall, K.R. 1978. The characteristics of effluents from New Zealand dairy factories. Bull. Ind. Dairy Fed. (104) 123-6.

Mathis, B.J. and Kavern, W.R. 1975. Distribution of mercury, cadmium, lead and thallium in a eutrophic lake. Hydrobiologia. 46 (2-3) 207-222.

Mathur, A., Sharma, Y.C. Rupainwar, D.C., Murthy, B.C., and Chandra, S. 1987. A study of river Ganga at Varanasi with special emphasis on. Heavy metal pollution. Poll. Res. (6) 37-44.

Matulewich, V.A. and Finstin, M.S. 1978. "Distribution of autotrophic nitrifying bacteria in a polluted river (the Passaic)". AppI. and Environ. Microbiol. (35)1.

Mayes, R.A., McIntosh, A.W., and Anderson, V.L. 1977. Uptake of cadmium and lead by a rooted aquatic macrophyte *(Elodea canadensis)*. Ecology. (58) 1176-1180.

Mc Donnel 1982. Oxygen budget in macrophyte impacted streams. Water Research. (16) 1037-1076.

McBay, L.G. 1961. The biology of *Tilapia aurea*. Proc. Annu. Conf. Southeast Assoc. Game Fish Comm. 15 208-218.

McDowall, F.H. and Thomas, R.H. 1961. Disposal of Dairy Wastes by Spray Irrigation on Pasture Land, Pollution Advisory Council Pub. No.8. Wellington, New Zealand.

McGarry, M. and C. Tongkasome. 1971. Water reclamation and algae harvesting. Water Control J. (43) 824-835.

McIntosh, A.W., Shephard, B.K. Mayer, R.A. Atchison, G.J. and Nelson, D.W. 1978. Some aspects of sediment distribution and macrophyte cycling of heavy metals in a contaminated lake. J. Environ Qual. (7) 301-305.

McLaren, et al. 1976. Effective phosphorus removal from sewage by biological means. Water SA. No.1.

McNabb, C.D. Jr. 1976. The potential of submersed vascular plants for reclamation of waste water in temperate zone ponds. In Tourbier, J. and R.W. Pierson, Jr. (eds.) Biological control of water pollution. University of Pennsylvania Press.

Mechsner, K. and Wuhrmann, K. 1974. Ecological considerations and an exemplification on biological treatment for dairy wastes. Dairy Effluent Treatment, Doc. 77, IDF, Brussels. 75-84.

Meenakshi, S. et al. 1991. The pH dependence of efficiency of activated alumina in defluoridation of water. Indian J. Env. Prot. (7) 511-513.

Mehta, I. and Sharma, R.K. 1972. Control of aquatic weeds by the white amur in Rajasthan, India. Hyacinth Control J. (10) 16-19.

Merchyulenene, D.P. and Nyanishkene, V.B. 1976. Accumulation of lead-210 by fresh water plants. Radiobiology. (16) 200 - 204.

Metcalf and Eddy, Inc. 1991. Waste water engineering Treatment, disposal and reuse. 3d ed. McGraw-Hill.

Middlebrooks, E.J. 1965. Taste and odour control in water. Public Works. 96 (4) 127.

Mikkelsen, D.S., S.K. De Datta, W.N. Obcemea 1978. Ammonia volatilization doses from flooded rice soils. Soil Sci. Sci. Am. I. (42) 725-730.

Miner, J., J.W. Wooten, and J.D. Dodd. 1971. Water hyacinths to further treat anaerobic lagoon effluent. Proceedings of International Symposium on Livestock Wastes 170-172.

Minshall, W.H. and Scarth, G.W. 1952. Effect of growth in acid media on the morphology, hydrogenation concentration, viscosity and permeability of water hyacinth and frogbit root cells. Canad. J. Bot. (32) 188-208.

Mishra, K. 1993. Ecology and Management of Freshwater Pond of Varanasi, Ph.D. Thesis, Banaras Hindu University, Varanasi, India.

Misra, R. 1946. A study in ecology of low lying lands. Indian Ecologist. (1) 1-20.

Misra, R. 1968. Ecology workbook. Oxford and IBH Publ. Co. Calcutta (29).

Misra, R.D. 1938. Edaphic factors in the distribution of aquatic plants in the English lakes, J. Ecol. (29) 280-329.

Mitchell, D.S. 1973. The growth and management of *Eichhornia crassipes* and *Salvinia spp.* in their native environment and in Alien situations. Aquatic weeds in S.E. Asia ed. C.K. Varshrey and J. Rozoska. Dr. W. Jurk B.V. Publishers, Netherlands. p.1 67-176.

Moorhead, K.K., K.R. Reddy and D.A. Graetz. 1988. Nitrogen transformations in a water hyacinth based water treatment system. J. Environ. Qual. Vol. 17 (1) 71-75.

Mortimer, D.C. 1985. Fresh water aquatic macrophytes as heavy metal monitor. The Ottowa river experience. Environmental monitoring and assessment. (5) 311-323.

Moshiri, G.A. (ed.) 1993. Constructed wetlands for water quality improvement. Lewis Pubi., Boca Raton, FL.

Mukherjee, D.P. 1982. Treatment of domestic sewage by aquaculture. Fertilizer Technol. 19(3 and4) 111-117.

Munshi, J.D., Dutta G.R., Singh N.K. 1989. Biological control of heavy metal pollution by some aquatic macrophytes in river Subernrekha. J. Fresh Wat. Biol. (1) 25-3 1.

Murphy, J. and Riley, J.P. 1962. A modified single solution method for the determination of phosphate in natural waters. Analyt. Chim. Acta. (27) 31-36.

Murthy, Y.S., V. Subbaiah, D.S. Rao, R.C. Reddy, L. Shanti Kumar, S.I. Elyas, K.G. RamaRao, J.S. Gadgil and S.B. Deshmukh. 1984. Treatment and disposal of waste waters from synthetic drugs plant (IDPL) Hyderabad; Part I Waste water characteristics. Indian J. Env. Health. 26(1).

Mutukumaran, K., B. Balusubramanian and T.V. Ramakrishna 1995. Removal of fluoride by chemically activated carbon. India J. Env. Prot. 15 (7) 514-517.

Myers R.W. 1977. A comparative study of nutrient composition and growth of selected duckweeds, Lemnaceae, on dairy waste lagoons California Polytech. State University.

Nalewajko, C. and D.R.S. Lean. 1980. Phosphorus. In the physiological, ecology of phytoplankton. Ed. I. Morris. Blackwell Scientific Publications. p.235-258.

Nandan, S.N. 1985. Eutrophication in Vishwamitri river flowing through Baroda city (India) as indicators of organic pollution. Ind. J. Ecol. 10(1) 11-15.

National Dairy Development Board (NDDB) 1993. Operation and maintenance of effluent treatment plant. Mansinh Inst. of Training, Mehsana, India.

Neel, J.K. 1963. Impact of reservoirs. pp. 575-594. In D.G. Fr. (ed.). Limnology in North America. Univ. Wisconsin Press, Madison.

Neeteson, J.J. 2000. Nitrogen and phosphorus management on Dutch dairy farms: legislation and strategies employed to meet the regulations. Biology and Fertility of Soils. 30 (5-6) 566.

Nemerow, N.L. 1976. Aguas residuales industriales. Madrid : Blume.

Newman, M.C. and McIntosh, A.W. 1983. Slow accumulation of lead from contaminated food sources by fresh water gastropods. *Physa integra* and *Campeloma decisum*. Arch. Environ. Contain. Toxical. (12) 685-692.

Nicholls, H.A. and P.W. Osborn. 1979. Bacterial stress : phosphorus. J. WPCF 51(4) 557.

Nielson, L.K. 1975. Seasonal variation in sediment water exchange of nutrientions in lake Erromvert, Int. Verein. Limnol. (19) 1057-1065.

Nielson, L.P., P.B. Christensen, N.P. Revsbech, and J. Sorensen. 1990. Denitrification and oxygen respiration in biofllms studied with a microsensor for nitrous oxide and oxygen. Microb. Ecol. (19) 63-72.

Nolan, W.J. and D.W. Kirmse. 1974. The paper making properties of water hyacinth. Hyacinth Control J. (12) 90-97.

Nommik, H. 1956. Investigations on denitrification in soil. Acta. Agric. Scand. VI. (2) 195-228.

North, W.J. (ed.) 1971. The biology of giant Kelp beds *(Macrocystis)* in California. J. Cramer Publishers. Lehre, Germany.

Obeid, M. and Chadwick, M.J. 1964. Some factors affecting the growth of two aquatic weed species of Nile-water hyacinth

(*Eichhornia crassipes* Solms.) and water lettuce (*Pistia stratiotes L.*) Proc. 7th Br. Weed Control Conf. 548-52.

Obeid, M. 1962. An investigation into the mineral nutrition of some common weed species in the Sudan. M.Sc. thesis, University of Khartoum.

Odum, E.P. 1971. Fundamentals of Ecology, W.B. Saunders Company, Philadelphia.

Odum, H.T. 1956. Primary production in flowing water. Limol. Oceanogr. (1) 102-117.

Ogwada, R.A., K.R. Reddy and D.A. Graetz. 1984. The effects of aeration and temperature on nutrient regeneration from selected aquatic macrophytes. J. Environ. Qual. (13) 239-243.

Olaniya, M.S., K.L. Saxena and H.C. Sharma. 1976. Pollution studies of Chambal river and its tributaries at Kota. Indian J. Environ. Hlth. 18(3) 2 19-226.

Oron, G., Vegtt, A. DE and Porath 1988. Nitrogen removal and conversion by duckweed grown on waste-water. Wat. Res. 2(22) 179-184.

Oswald, W. 1976. Gas production from microalgae. Proceedings of clean fuels from biomass, sewage, urban refuse, and agriculture wastes. 173-187.

Oswald, W.I., H.B. Gotaas, C.G. Golueke, and W.R. Kellen. 1957. Algae in waste treatment. Sewage Ind. Wastes (29) 437-455.

Outridge, P.M. 1992. Comparing cadmium toxicity test with plants in monocultures and species mixtures. Bull. Environ. Contain. Toxicol. (48) 344-351.

Ovon, G., Porath, D. and Wildschut, I.R. 1936. Waste water treatment and renovation by different duckweed species. J. Environ. Engineers. (112) 247-63.

Palmela, M. 1986. Ecological effects of acidification on primary producers in aquatic system. Water Air and Soil Poll. (30) 421-438.

Palmer, C.M. 1969. A composite rating of algal tolerating organic pollution. J. Phycol. (5) 78-82.

Pano, A. and Middlebrooks, E.J. 1982. Jour. Water Pollut. Control Fed. 54 (4) 344-351.

Parent, L., Allard, M., Planas, D., and Moreau, G. 1985. The effects of short term and continuous experimental acidification on biomass and productivity of running water periphytic algae, Proc. ASTM Symposium.

Parham, B.E.V. 1947. Weed control studies in Fiji. II Eradication of water hyacinth and other aquatic weeds. Agric. J. Dep. Agric. Fiji. (18) 3 5-42.

Parija, P. 1934a. Physiological investigations on water hyacinth *(Eichhornia crassipes)* in Orissa with notes on some other aquatic weeds. Indian J. Agric. Sci. (4) 399-429.

Parija, P. 1934b. A note on the reappearance of water hyacinth seedlings in cleared tanks. Indian J. Agric. Sci. (4) 1049.

Parkin, M.F. and Marshall, K.R. 1976. Spray irrigation disposal of dairy factory effluent — a review of current practice in New Zealand. N.Z. J. Dairy, Sci. Technol. (11) 196-205.

Parra, J.V. and C.C. Hortenstein. 1974. Plant nutritional content of some Florida water hyacinths and response by pearl millet to incorporation of water hyacinths in three soil types. Hyacinth Control J. (12) 85-90.

Patemiti, M.B. and K.E. Mantai. 1986. Shoot/leaf uptake of phosphate by *Myriophyllum spicatum* and *Ceratophyllum demersum.* Proc. 7th International Symp, on Aquatic Weeds. 239-244.

Paul, L., Bishop, J. and Eighmy, T. 1989. Aquatic water treatment using *Elodea nuttall* ii. Jour. Wat. Poll. Cont. Fed. 6 1(5) 64 1-648.

Pavoni, S., Sfriso, R., Donazzolo, R. and Drio, A.A. 1990. Influence of waste water from the city of Venice, Italy and the Hinterland on the eutrophication of the lagoon. Sci. Total Envi. 96 (3) 235-252.

Peach, K. and M.V. Tracey. 1956. Modern methods of plant analysis. Vol.1. Springers-Verlag, Berlin.

Pearl, H.W. and N.D. Bowles. 1987. Dilution bioassays their application to assessments of nutrient limitation in hypereutrophic waters, phytoplankton, algal bloom, water quality growth. Neuse Rive North. Carolina, U.S.A. Hydrobiologia. 146 (3) 265-274.

Peckol, P.B., B. Demeo-Anderson, J. Rivers, I. Valiela, M. Maldonado, and J. Yates. 1994. Growth nutrient uptake capacities and tissue constituents of the macroalgae *Cladophora vagabunda* and *Geaciria tikvahiae* related to site-specific nitrogen loading rates. Mar. Biol. (121) 175-185.

Penfound, W.T. 1956. Primary production of vascular aquatic plants. Limnology and Oceanography (1) 92-101.

Penfound, W.T. and T.T. Earle. 1948. The biology of the water hyacinth. Ecol. Monogr. (18) 447-472.

Peters, G.A., R.E. Toia, W.R. Evans, D.K. Crist B.C. Mayne, and R.E. Poole. 1980. Characterization and comparison of five N fixing Azalla-Anabaena associations I. Optimization of growth conditions for biomass increase and N content in a controlled environment. Plant Cell Environ. (3) 26 1-269.

Philip, C.B. 1927. Diurnal fluctuations in the hydrogen ion and productivity of a Minnessta lake. Ecology (8) 73-89.

Piccardi, E.B. and Clauser, M. 1983. Absorption of copper by *Iris pseudocaris.* Water air and Soil Poll. (19) 185-192.

Pip, E. 1987. The ecology of *Potamogeton* sp. in Central North America. Hydrobiol. (153) 203-2 16.

Pip, E. 1990. Cadmium, copper and lead in aquatic macrophytes in Shoal Lake (Manitobaontario). Hydrobiol. (208) 253-260.

Pip, Eva. 1979. Survey of the ecology of subnierged aquatic macrophytes in central Canada. Aquatic Bot. (7) 339-357.

Pine, N.W. 1960. Water Hyacinth a curse or a crop? Nature. (185) 116.

Planas D., S.C. Mabenly, and J.E. Parker. 1996. Phosphorus and nitrogen relationships of *Cladophora glomerata* in two lake basins of different trophic status. Freshwater Biol. (35) 609-622.

Porges, N., Michener, T.S. Jr., Jasewicz, L. and Hoover S.R. 1960. Dairy waste Treatment toy Aeration Theory, Design, Construction and Operation, Agr. Handbook 176, Agr. Research Service, Washington, DC.

Prakasham, R.S., Reddy L. Poorna Chandra, Amita, Manisha and Ramakrishna S.V. 1998. Defluoridation of drinking water using *Eichhornia* sp.

Pritchard, P.M. 1985. Fate and effect of pollutants Fate of environment pollutants. J. Wat. Poll. Cont. Fed. 57 (6) 658-662.

Rai, D.N. and Datta Munshi, J.S. 1977. Observations on diurnal fluctuations of certain physico-chemical factors of three tropical swamps of Darbhanga (India). Comp. Physiol. and Ecol. Vol.3.

Rai, D.N. and Datta Munshi. 1979. Influence of thick floating vegetation (water hyacinth *Eichhornia crassipes*) on the physico-chemical environment of a fresh water wetland. Hydrobiologia. 62 (1) 65-69.

Rai, H. and G. Hill 1981a. Physical and chemical studies of Lago type. A control Amazonian Black Water, Ria Lake Int. Revueges. Hydrobiol. 66(1) 37-82.

Rai, H. and G. Hill 1982. On the nature of the ecological cycle of Lago Janauari. A central Amazonian Ria Varzea Lake. Tropical Ecology. 23(1) 1-49.

Rai, H. and G. Hill. 1981b. Bacterial biodynamics in Lago type. A control Amazonian Black Water. Ria Lake. Arch. Hydrobiol. Supp. 54(4) 420-468.

Rai, H. and G. Hill. 1984. Primary production in the Amazonian aquatic ecosystem. The Amazon (Ed. By H. Sioli). 311-335.

Rai, L.C. and Mallick, N. 1992. Removal and assessment of toxicity of Cu and Fe to Anabaena daliolum and Chiorella vulgaris using free and immobilized cells. World J. Microbiol. Biotechnol. (8) 110-114.

Rai, U.N. and Chandra, P. 1989. Removal of heavy metals from polluted waters by Hydrodictyon reticulatum (Linn.) Lagerheim. Sd. Total Env. (87/88) 509- 15.

Rai, U.N. and Chandra, P. 1992. Accumulation of Cu, Pb, Mn and Fe by field population of Hydrodictyon reticulatum (Linn.) Lagerheim. Sci. Total Env. (116) 203-211.

Rai, U.N., Sinha, S., Tripathi, R.D., Chandra, P. 1994. Waste water treatability potential of some aquatic macrophytes Removal of heavy metals. Ecolog. Enge.

Raina, V., A.R. Shah and R. Ahmad Shakto 1984. Pollution studies on water quality. Ind. Jr. Environ. Hlth. 26 (3) 187-201.

Rajgopalan, S., A.K. Basu, R.S. Dhaneshwar and C.S.G. Rao 1970. Pollution of river Subamrekha at Ranchi – A survey. Env. Hith. (12) 246-259.

Rajgopalan, S., G.J. Mohan Rao, D. Raghuraman, M.K. Abdullappa, A.C. Manuel and C.G. Mehta. 1973. Pollution of Mahi river. Env. HIth. 15(2) 100-110.

Rajsekhar, T., Shah, D.D., Prajapati, P.S. and Joshi, D.C. 2000. Mechanical aerators for dairy effluent treatment plant. Indian Dairyman. (52) 4.

Rao and Mathur, S.N. 1975. Chemical eradication of Eichhornia. Proc. Ind. Nati. Sci. Acad.

Rao, D.S., R.C. Reddy, V. Subbaiah, Y.S. Murthy, K.G. Rama Rao and K. Annapurna. 1985. Treatment and disposal of waste waters from organo-phosphorous pesticides factory. Presented at National Seminar on pollution control and environmental management; Vol II : Waste treatment, recycling, reuse and disposal. NEERI, Nagpur.

Raschke, R.L. 1968. Algal periodicity, primary production, and waste reclamation in a tertiary sewage stabilization pond ecosystem. Unpublished Ph.D. thesis, Iowa State Univ. Ames.

Raschke, R.L. 1970. Algal periodicity, and waste reclamation in a stabilization pond ecosystem. JWPCF 42(4) 518- 530.

Raveh, A., and Y. Avnimelech. 1973. Minimizing nitrate seepage from the Hula valley into Lake Kinneret (Sea of Galilee). I. Enhancement of nitrate reduction by sprinkling and flooding. J. Environ. Qual. (2) 455-458.

Reddy, K.R. 1981. Diel variations in physico-chemical parameters of water in selected aquatic systems. Hydrobiologia. (85) 201-207.

Reddy, K.R. 1982. Nitrogen and phosphorus cycling in shallow reservoirs used for agricultural drainage water treatment. In Symp. Nutrient Cycling in Agri. Ecosys. Ann Arbor, Mich. Univ. Ga., Athens (in press).

Reddy, K.R. 1983. Fate of nitrogen and phosphorus in a waste water retention reservoir containing aquatic macrophytes. J. Environ. Qual. (12) 137-141.

Reddy, K.R. and D.A. Graetz. 1981. Use of shallow reservoir and flooded soil systems for waste water treatment, nitrogen and phosphorus transformation. J. Environ. Qual. (10) 113-119.

Reddy, K.R., and W.F. DeBusk. 1984. Growth characteristics of aquatic macrophytes cultured in nutrient enriched water :1. Water hyacinth, water lettuce, and pennywort. Econ. Bot. (38) 225-23 5.

Reddy, K.R., and W.F. DeBusk. 1985. Growth characteristics of aquatic macrophytes cultured in nutrient enriched water :11 AzoiZa, duckweed, and Sqivinia. Econ. Bot. (39) 200-208.

Reddy, K.R., K.L. Campbell, D.A. Graetz and K.M. Portier. 1982. Use .of biological filters for agricultural drainage water treatment. J. Environ. Qual. (11) 591-595.

Reddy, K.R., P.D. Sacco, D.A. Graetz, K.L. Campbell and L.R. Sinclair. 1982. Water treatment by an aquatic ecosystem

Nutrient removal by reservoirs and flooded fields. Environ. Mgt. (6) 261-271.

Reddy, K.R., Sacco, P.D. and D.A. Graetz. 1980. Nitrate reduction in an organic soil-water system. J. Environ. Qual. 2(9) 283-288.

Reddy, K.R., Sacco, P.D., Graetz, D.A., Campbell, K.L. and Sinclair, L.R. 1981. Aquatic ecosystems as a means of agricultural drainage water treatment efficiency of nutrient removal by reservoirs and flooded fields. Environmental Management.

Redshaw, C.J., Mason, C.F., Haynes, C.R. and Roberts, R.D. 1988. Nutrients budget for a hypertrophic reservoir. Water Research. 22 (4) 413-419.

Riddle, M.J. and Chandler, W.D. 1974. Waste disposal of whey. Proc. Whey Utilization Symposium, FM, DD, Canada Department of Agriculture. 94-123.

Rodrignes, A.M. and Oliveira, J.F.S. 1987. Treatment of waste water from the tomato concentrate industry in high rate algal ponds. Wat. Sa. Tech. (19) 43-9.

Rogers, H.H. and D.E. Davis. 1972. Nutrient removal by water hyacinths. Weed Sci. (20) 423-42 8.

Rosen, A..A.., J.B. Peter and F.M. Middleton. 1962. Odour thresholds of mixed organic chemicals. J. Water Poll. Cont. Fed. (34) 7.

Rosen, A.A., R.T. Skeel and M.B. Effinger. 1963. Relationship of river water odour to specific organic contaminants. J. Water Poll. Cont. Fed. (35) 777.

Roush, T.H. 1985. Fate and effects of pollutants. Effect of pollution on fresh water organisms.. J. Water Poll. Control. Fed. 57(6) 167-171.

Royal, L. 1978. Reduction of milk and milk product wastage, Proc. IDF Seminar on Dairy Effluents. Doc. 104, IDF. Brussels. 17-27.

Ruschel, A.P. 1987. N uptake by Azolla-Anabaena Plant and Soil. 97(1) 79-83.

Ryther, J.H., and W.M. Dunstan. 1971. Nitrogen, phosphorus and eutrophication in the coastal marine environment. Science. 1(71) 1008-1013.

Salisbury, F.B. and Ross, C.W. 1995. Plant Physiology CBS Publishers and Distributors, Delhi, India.

Sammaiah P. and Sastry C.A. 1991. Dairy Waste Water Treatment using Rotating Biological Disc Contactor. Indian J. Environ. Prot. 11(5) 34 1-346.

Sanjay Kumar 1995. Studies on desorption of fluoride from activated alumina. Indian J. Env. Prot. 15 (1) 50-53.

Santos, M.C.R. and Oliveira, J.F.S. 1987. Water Sci. Technol. 19 (12) 123-130.

Sarkar, A. and Jana, 5. 1986. Heavy metal pollutant tolerance of Azolla pinnata. Water, Air and Soil Pollution. (27) 15-18.

Sastry, C.A. 1995. Control of aquatic weeds from water bodies using grass carp. Indian J. Env. Health. 15(2) 92-99.

Sawyer, C.N. and P.L. McCarty. 1978. Chemistry for environmental engineering (3rd edn.). McGraw-Hill Kogakushia Ltd., Tokyo, Japan.

Sayed, M.S. 1997. Pollution control in dairy plant. Dairy India. p.351.

Scarsbrook, B. and D.E. Davis. 1971. Effect of sewage effluent on growth of five vascular aquatic species. Hyacinth Control J. (9) 26-30.

Schraufnagel, F.H. 1957. Dairy wastes disposal by ridge and furrow irrigation, th Proc. 12 Ind. Waste Conf., Purdue University, Indiana. 28-49.

Schuch, R., R. Gensicke, K. Merkel, J. Winter 2000. Nitrogen and DOC removal from waste water streams of the metal working industry. Water research. 34 (1) p.295.

Schulze, K.L. 1966. Biological recovery of waste water. J. Water Poll. Control. Fed. (38) 1944-1958.

Sculthorpe, C.D. 1967. The Biology of Aquatic Vascular Plants. Edward Arnold (Publishers) Ltd. London.

Seethapathy Rao 1964. Defluoridation of water using sulphonated coconut shell carbon. Indian J. Env. Health. 6(1) 11-12.

Seidel, K. 1971. Macrophytes as functional element in the environment of man. Hydrobiologia. (12) 121-130.

Seidel, K. 1976. Macrophytes and water purification. In Tourbier, J. and R.W. Pierson, Jr. (eds.). Biological control of water pollution. University of Pennsylvania Press.

Sela, M., Gaity, J. and Tel-or, E. 1989. The accumulation and the effect of heavy metals on the water fern Azolla fihiculoides. New Phytol. (122) 712.

Selvapathy P.C., Sreedhar, P. 1991. Heavy metal removal by water hyacinth. J. Instr. Pubi. Hith. Eng., India. (3) 11-17.

Seyfried, C.F. 1974. Treatment of dairy effluent by plastic medium trickling filters. Dairy Effluent Treatment, Doc. 77, IDF, Brussels. 10 1-107.

Shaker, N., Abo-donia, S., Abd-el Shaheed, Y. and Ismail, A. 1988. Effect of lactic acid bacteria and heat treatments on pesticides contaminated milk. Egyptian J. Dairy Sci. 16(2) 309-3 17.

Sharma, B.K. 2000. Environmental Chemistry. Goel Publishing House, Krishna Prakashan Media (P) Ltd., Meerut-250001, India.

Shelef, G., Azov, Y., Moraine, R. and Oron, G. 1980. Algae mass production as an integral part of a waste water treatment and reclamation system. In Algae Biomass Production and Use, ed. G. Shelef and C.J. Soader, Elsevier North Holland Biomedical Press, Amsterdam, pp. 163-89.

Shiralipour, A. and P.H. Smith. 1984. Conversion of biomass into methane gas. Biomass. (6) 85-94.

Shukla, S.C. and Tripathi, B.D. 1989. Biological treatment of domestic waste water by water hyacinth and algal culture. Science and Culture. (55) 209-11.

Sikandar, Mohd. 1986. Ecology of river Ganga in Varanasi with special reference to pollution. Ph.D. Thesis. Banaras Hindu University, Varanasi.

Silvey, J.K.G., R. Allison, R.C. Hoehn and R. Cates. 1965. Taste and odour control in water: 4 parts. Southwest Water Works J. (47) 19, 22, 26, 32.

Singh, A.K. and B.N. Bhowmick. 1985. Effect of sewage on physicochemical characteristics and bacterial population of river Ganga at Patna. Ind. J. Ecol. 12(1) 17-19.

Singh, S.N. and Bhargava, A.K. 1985. Plant distribution as effected by the confluence of a polluted stream with a clean stream. Ind. J. of Ecol. (1) 15-19.

Sinha, S., U.N. Rai, R.D. Tripathi and P. Chandra. 1993. Chromium and manganese uptake by *Hydrilla verticillata* (1 .f.) Royle : amelioration of chromium toxicity by manganese. J. Env. Sci. Health. (A 28) 1545-1552.

Sinicorpe, T.L., Langio R., Gersberg R.M., Busnardo M., Zedler J. 1992. Metal removal by wetland mesocosms subjected to different hydro periods. Ecolog. Enge. (1) 309-322.

Snedecor, G.W. and Cochran, W.G. 1989. Statistical Methods Iowa State University Press, Ames, Iowa. 5001Q

Sorensen, J. and N.P. Revsbech. 1990. Denitrification in stream biofils and sediment In situ variation and control factors. In N.P. Revstrech and J. Sorensen (ed.) Denitrification in soil and sediment. Plenum Press, New York.

Sorensen, J., T. Jorgensen, and S. Brandt. 1988. Denitrification in stream epilithon : Seasonal variation in Gelback and Rabisback, Denmark. FEMS Microbiol. Ecol. (53) 345-354.

Spector, M.L. 1977. Production of non-bulking activated sludge. U.S. Patent 4 (056) 465.

Spencer, D.F., Ksander, G.G. (Davis, CA, USA), Madsen, JD. (Vicksburg, MS., USA) and Owens, C.S. (Lewisville, TX, USA). 2000. Aquatic Botany. (67) 23 7-240. Elsevier No.3.

Sreenivasan, A., Soundararaj, R. and Franklin, T. 1975. Diurnal and seasonal changes in a productive shallow tropical pond. Phykos. (12) 76-103.

Srivastava, V.C. and Sahai, R. 1976. Effect of water pollution on productivity and periodicity of phytoplankton of Chilwa lake. Geobios. 2 (6):187-190.

Standard methods for examination of water and wastewater. 1995. American Public Health Association, American Water Works Association, and Water Pollution Control Federation, Washington, D.C.

Stanford, G., R.A. Vanderpol, and S. Dzienia. 1975a. Denitrification rate in relation to total and extractable soil carbon. Soil Sci. Soc. Am. Proc. (39) 284-289.

Stanford, G.S. Dzienia, and R.A. Vanderpol. 1975b. Effect of temperatures on denitrification rate in soils. Soil Sci. Soc. Am. Proc. (39) 867-870.

Staves, R.P. and Knaus R.M. 1985. Chromium removal by three species of duckweeds. Aquat. Bot. (23) 26 1-263.

Stensel, H.D., R.C. Loehr, and A.W. Lawrence. 1973. Biological kinetics of suspended growth denitrification. 3. Water Pollut. Control. Fed. (45) 249-261.

Sterkey, J.E. and Karr, P.K. 1984. Effect of low dissolved oxygen concentration of different turbidity. 3. Wat. Poll. Cont. Fed. 56 (7) 83 7-843.

Steward, K.K. 1970. Nutrient removal potentials of various aquatic plants. Hyacinth Control 3. (8) 34.

Stowell, R.M., R. Ludwig, J. Colt and G. Tchobanoglous. 1981. Concepts in aquatic treatment system design. ASCE 3. Environ. Eng. Div. (107) 9 19-940.

Stratton, F.E. 1969. Nitrogen losses from alkaline water impoundments. ASCE J. Sanit. Eng. Div. (95) 223-231.

Strom, A. 1974. Some parameters expressing the pollution of dairy effluent, Dairy Effluent Treatment, Doc. 77, IDF, Brussels. 18-23.

Sutton, D.L. and Blackburn, B.D. '1971. Uptake of copper by water hyacinth. Hyacinth Control J. (9) 18-20.

Sutton, D.L. and W.H. Ornes. 1975. Phosphorus removal from static sewage effluent using duckweed. J. Environ. Qual. (4) 367-370.

Sutton, D.L. and W.H. Omes. 1977. Growth of *Spirodela polyrhiza* in static sewage effluent. Aquatic Botany (3) 231-237.

Svensson, J. 1993. Denitrification measurements in the Kallby ponds on 14 August 1993. Dep. of Limnology, Lund University, Lund (In Swedish).

Svoboda, M. 1974. Waste stabilization ponds and waste stabilization basins, Dairy Effluent Treatment, Doc. 77, IDF, Brussels. 85-92.

Svoboda, M., Gillar, J., Hlavaka, M., Salplachta, J., Stelcova, D. and Marvan, P. 1966. Purification of dairy wastes by means of lagoons XVII Int. Dairy Cong. ElF. 7 15-722.

Symons, G.E. 1956. Taste and odours Part 1 and 2. Water Sewerage Works. (103) 307-348.

Synnott, E.C., Kelly, B.F. and Moloney, A.M. 1978. Recent developments in dairy effluent treatment in Ireland. Proc. IDF Seminar on Dairy Effluents, Doc. 104, IDF, Brussels. 156-159.

Tam, N.F.Y. and Wong, Y.S. 1989. Waste water nutrient removal by *Chlorella pyrenoidosa* and *Scenedesmus* sp. Environ. Pollut. (58) 19-34.

Tchobanoglous, G. 1987. Aquatic plant systems for waste water treatment Engineering considerations. Aquatic plants for water treatment and resource recovery, K.R. Reddy and W.H. Smith, eds., Magnolia Publishing, Orlando, Fla. (27) 48.

Tchobanoglous, G., Maitski, F., Thompson, K., and Chadwick, T.H. 1989. Evolution and performance of city of San Diego pilot-scale aquatic waste water treatment system using water hyacinths. Res. J. Water Pollution Control Fedn. 61(11/12), 1625-1635.

Tecator. 1992. Determination of total Kjeldahl nitrogen by flow injection analysis. Application Note. ASN 134 0 1/92. Tecator, Hoganas.

Thakral, S.K., G.B. McPherson, M. Iverson, and G. Lai. 1994. Use of constructed wetlands for effluent denitrification and environmental enhancement A win-win solution for the city of Riverside, Calif. pp. 91-99. In Proc. 67th Annual Conf. and Exposition. Water Environment Federation, Chicago, IL. 15-19 Oct. 1994. Water Environment Federation, Alexandria, V.A.

Tiedje, J.M. 1982. Denitrification. pp. 1011-1026. In Al. Page et al. (ed.) Methods of soil analysis. Part 2. Agron. Monogr. 9. ASA and SSSA, Madison, WI.

Tiedje, J.M. 1988. Ecology of denitrification and dissimilatory nitrate reduction to ammonium. In Zehnder, A.J.B. (Ed.) Biology of Anoerbic Micro-organisms. John Wiley & Sons, New York.

Timperley, M.H. 1978. Collaborative tests of water analysis. The CHEMAQUA Programme 3. Trace metals, N.Z. Jl. Sci. (21) 557-564.

Tiwari, T.N., S.C. Das and R.K. Bose. 1986. A relation between COD and BOD for the Ganga at Kanpur. Indian J. Env. Prot. 6(3) 183-184.

Toms, I.P., Owens, M., Hall, J.A. and Mindenhall, B.A. 1975. Water Pollut. Control Fed. (47) 383-401.

Tracy, K.D., and A. Flammino. 1987. Biochemistry and energetics of biological phosphorus removal. Pergamon Press.

Trebler, H.A. and Harding, H.G. 1955. Fundamentals of the control and treatment of dairy waste. Sew. Ind. Wastes. (27) 1369-13 82.

Tripathi, B.D. 1986. Recent advances in pollution research in India. In recent Advances in Environmental Biology. R.S. Ambasht (Ed.) Professor D.N. Rao Commemoration Volume. 72-79.

Tripathi, B.D. .and S.C. Shukla. 1991. Biological treatment of waste water by selected aquatic plants. Environmental Pollution. (69) 69-78.

Tripathi, B.D., J. Srivastava and K. Misra. 1990. Impact of pollution on the elemental composition of water hyacinth *Eichhornia crassipes* (Mart, Soims.) and Lemna *(Lemna minor L.)* in various ponds of Varanasi. Science and Culture. (55) 30 1-8.

Tripathi, B.D., R.S. Ambasht and Mohd. Sikandar. 1984. Ecological investigation of energy consumption for burning of dead bodies at Varanasi Ganga ghats. pp. 1 In R.S. Ambasht and B.D. Tripathi (eds.). River Ecology and Human Health. NECA, Varanasi.

Tripathi, B.D., Srivastava, J. and Misra, K. 1990. Impact of pollution on the elemental composition of water hyacinth [*Eichhornia crassipes* (Mart.) Solms] and Lemna *(Lemna minor L.)* in various ponds of Varanasi. Science and Culture. (55) pp. 301-8.

Tripathi, B .D., Srivastava, J. and Misra, K. 1991. Nitrogen and phosphorus removal-capacity of four-chosen aquatic macrophytes in tropical fresh water ponds. Environ. Conser. 18(2) 143-47.

Tripathi, B.D., Upadhyay, A.R., Dwivedi, U.P., Singh, M.K., Dwivedi, A.K., Pandey, T. and Pandey, A.K. 1998. Nitrogen and phosphorus contents of water, sediment and some aquatic macrophytes of Ratoi Taal; Distt. Mau, U.P., India. Adv. Biol. Res. 16 (2) 1-7.

Tripathi, B.D., Upadhyay, A.R., Dwivedi, U.P., Singh, M.K., Dwivedi, A.K., Pandey, T. and Pandey, A.K. 2000. Characterization of effluent of a feeder balancing dairy at Varanasi. Adv. Biol. Res. 18(1) 1-10.

Tripathi, G. 1983. Environmental problems of Indian rivers and their ill effects. River Pollution and Human Health. R.S. Ambasht and B.D. Tripathi (Eds.) NECA, Varanasi, 23-33.

Tripathi, R.D. and Chandra, P. 1991. Chromium uptake by Spirodela polyrrhiza (L.) Schleiden in relation to metal chelators and pH. Bull. Environ. Contain. Toxicol. (47) 764-769.

Tucker, S.C., and T.A. DeBusk. 1981. Productivity and nutritive value of *Pistia stratiotes* and *Eichhornia crassipes*. J. Aquatic Pl. Mgt. (19) 61-63.

Tukunga, T., Furstas, N. and Morimoto, M. 1976. Accumulation of Cd in *E. crassipes*. Eiseikaguka. (22) 234-239.

Tuszynski, W.B. 1978. Packaging, Storage and Distribution of Processed Milk Technical Requirements and their Economic Implications, FAO Animal Production and Health Paper No.11, FAQ, Rome, Italy.

U.P. Pollution Control Board (U.P.P.C.B.). 1998. Dairy Industry-Effluent Quality. Lucknow, India.

U.S. Environmental Protection Agency (EPA). 1975. Process design manual for nitrogen control.

U.S. Public Health Service. 1962. Drinking water standards, PHS. Pub. 956. US Department of Health Education and Welfare, Washington DC.

Ultsch, G.R. and Anthony, D.S. 1973. The role of the aquatic exchange of carbondioxide in the ecology of the water hyacinth *(Eichhornia crassipes)*. Florida Sci. 36 (1) 16-22.

Upadhyay Alka, R. and B.D. Tripathi. 2001. Nutrient removal from dairy wastewater using aquatic macrophytes. Nat. Sem. on Emerg. Tren. in Environ. Manag. Deptt. of Botany, B.H.U., Varanasi.

Upadhyay, Alka. 1996. Dairy Industry, Waste Management and Effluent Treatment. M.Sc. Diss., University of Lucknow, Lucknow, India.

Van Haandel, A.C., Ekema, G.A. and G.V.R. Marais. 1981. The activated sludge process - 3. Water Res. (15) 1135-1152.

Vandamme, K. and Waes, G. 1980. Purification of dairy waste water in a two stage treatment plant including anaerobic pretreatment, Milchwissensch. (35) 663-666.

Vetter, R. 1972. Preliminary tests on the feeding value for cattle of fresh and processed water hyacinths. Iowa Agriculture and Home Economics Experiment Station Animal Science Leaflet R. (169) 1-3.

Vighi, M. and G. Chiaudhan. 1985. The impact of agricultural loads on eutrophication in EEC surface waters (Edited by Winteringham, F.P.W.) Backing, Essex, U.K. Elsevier Applied Science Publishers. 7 1-85.

Vollenweider, R.A. 1969. A manual on methods of measuring primary production in aquatic environments. Blackwell Science Publications, Oxford. p.213.

W.H.O. 1971. World Health Organization International standards for drinking rd water 3 . Edition, Geneva.

Wahlquist, H. 1972. Production of water hyacinths and resulting water quality in earthen ponds. Hyacinth Contr. J. (10) 9-11.

Wang, W. 1986. Toxicity tests of aquatic pollutants by using common duckweed. Environ. Pollut. (11) 1-14.

Warren, C.E. 1971. Biology of Water Pollution Control. W.B. Saunder's Co. Philadelphia.

Water Pollution Control Federation. 1983. Nutrient control manual of practice No. FD-7. Facilities Design.

Watson, K.S., Peterson, A.E. and Powell, R.D. 1977. Benefits of spreading whey on agricultural land, J. Water Pollut. Control Fed. (49) 24-34.

Weber, A.S., and Tchobanoglous, G. 1985. Nitrification in water hyacinth treatment systems. J. Environ. Engrg., ASCE. 111(5) 699-7 13.

Wells, J.M. & Brown, D.H. 1987. Factors affecting the kinetic of intra & extra cellular cadmium uptake by the moss *Rhytidiadeiphus squarrosus.* New Phytol. (105) 123-137.

Wells, J.M. & Brown, D.H. 1990. Ionic control of intracellular and extracellular Cd uptake by the moss the *Rhytidiadeiphus squarrosus.* New Phytol. (116) 54 1-553.

Wells, N.W. 1969. Differences in phosphate uptake rates exhibited by activated sludge. J. WPCF. 4 1(5) 765.

Westlake, D.F. 1963 Comparison of plant productivity. Biological Review. (38) 385-425.

Wetzel, R.G. 1983. Limnology (2nd edn.). Saunders.

Wheatland, A.B. 1974. Treatment of waste waters from dairies and dairy product factories methods and systems, J. Soc. Dairy Technol. (27) 7 1-79.

Wild, H.E., Sawyer, C.N. and McMohan, T.C. 1971. Jour. Wat. Pollut. Contr. Fed. (43) 1845-1854.

Wolverton, B.C. 1980. Higher plants for recycling human waste into food, potable water and revitalized air in a closed life support system. NASA/ERL Report No. 192, NSTL, MS.

Wolverton, B.C. 1981. Water hyacinth for controlling water pollution. In Varshney C.K. (ed.) Water Pollution and Management Reviews. South Asian Publisher, New Delhi, India. (47) 49.

Wolverton, B.C. 1982. Hybrid waste water micro-organisms and reed *(Pharagmites communis)*. Econ. Bot. (36) 373-380.

Wolverton, B.C. and Mc Donald, R.C. 1981. Energy of vascular plants used in waste water treatment systems. Economic Bot. (35) 224-232.

Wolverton, B.C. and McDonald, R.C. 1981. Natural processes for treatment of organic chemical waste. Environ. Profess. (3) 99-104.

Wolverton, B.C. and R.C. McDonald. 1976. Water hyacinths for upgrading sewage lagoons to meet advanced waste water treatment standards part II NASA Technical Memorandum TM-X-72730.

Wolverton, B.C. and R.C. McDonald. 1979. Water hyacinth *(Eichhor nia crassipes)* productivity and harvesting studies. Econ. Bot. (33) 1-10.

Wolverton, B.C., R.C. McDonald and J. Gordon. 1975. Water hyacinth and Alligator weeds for final filtration of sewage NASA Tech. Memo. TM-X-72724.

Wolverton, B.C., R.M. Barlow and R.C. McDonald. 1976. Application of vascular aquatic plants for pollution removal, energy and food production. In Tourbier J. and R.W. Pierson,

Jr. (eds.). Biological control of water pollution. University of Pennsylvania Press.

Wong, P.T.S. Mayfield, C.I. and Chau, Y.K.1980. Cadmium toxicity to phytoplankton and microorganisms.-In: Nariagu J.O. (ed.): Cadmium in the environment, Part I- John Wiley and Sons, New York. p.572-585.

Wooten, J.W. and J.D. Dodd. 1976. Growth of water hyacinths in treated sewage effluent. Econ. Bot. (30) 29-37.

Wuhrman K., 1964. Nitrogen removal in sewage treatment processes. Verh. Int. Verein. Limnol. (15) 580-96.

Wunderlich, W.E. 1967. The use of machinery in the control of aquatic vegetation. Hyacinth Control J. (6) 22-24.

Wunderlich, W.E. 1968. Aquatic plant control and the dollar. Hyacinth Control J. (7) 28-29.

Yamane, I. 1957. Nitrate reduction and denitrification in flooded soils. Soil Plant Food. (3) 100-103.

Yeoh, B.G. 1979. A preliminary study in the use of water hyacinth in the removal of heavy metal and BOD from industrial waste water. Conference on chemical research in Malaysia, Kualalumpur.

Yeoh, B.G. 1983. Use of water hyacinth in waste water treatment. Technical report. Standards of Industrial Research Institute (SIRIM) Res. Div. Malaysia. 67.

Yoshimuro, 5. 1932. Seasonal variation in the control of nitrogenous and phosphate in the water of Takasuka Pond. Saitma, Japan, Arch. Hydrobio. (24) 155-176.

Yoshizaki, S., T. Tomida. 2000. Principle and process of heavy metal removal from sewage sludge. Environmental Science and Technology. 34 (8) 1572.

Yount, James L., and R.A. Crossman, Jr. 1970. Eutrophication control by plant harvesting. J.W.P.C.F. (42) 173-183.

Zand, P.M. 1976. Indexes associated with information theory in water quality. J. Water Poll. Con. Fed. (48) 2026-2030.

Zirschky, J., Reed, S.C.. 1988. The use of duckweed for waste water treatment. J. Wat. Pollut. Cont. Fed. (60) 1253-1258.

INDEX